PRODUCE YOUR OWN FOOD 2-IN-1 COLLECTION

BACKYARD CHICKENS AND CONTAINER GARDENING—RAISE HAPPY HENS AND GROW YOUR OWN VEGGIES, HERBS, AND FLOWERS WITH EASE

AVERY SAGE

VIBRANT CIRCLE BOOKS LLC

CONTENTS

THE ULTIMATE BEGINNER'S GUIDE TO CONTAINER GARDENING

THE ULTIMATE BEGINNER'S GUIDE TO RAISING CHICKENS

THE ULTIMATE BEGINNER'S GUIDE TO CONTAINER GARDENING

SMALL SPACE GARDENING MADE EASY —GROW YOUR OWN VEGETABLES, HERBS, AND FLOWERS EVEN IF YOU HAVE NO EXPERIENCE!

INTRODUCTION

A few years ago, I met a neighbor named Sarah. She lived in a tiny apartment with only a small balcony. Sarah had never gardened before but was determined to bring some greenery into her life. Armed with a few pots, some soil, and a handful of seeds, she transformed her little outdoor space into a vibrant container garden. Her tomatoes flourished, her herbs thrived, and colorful flowers brightened her mornings. Watching her joy, I realized how powerful container gardening could be for anyone with limited space.

Container gardening is more than a trend; it's a growing movement. In cities worldwide, people are turning their balconies and windowsills into small, productive gardens. According to recent studies, urban gardening has increased by over 30% in the past decade. Many people want fresh produce but lack the yard space to grow it. Container gardening offers a solution, allowing you to cultivate your own food and enjoy nature, even in compact spaces.

The purpose of this book is simple. I aim to provide step-by-step guidance for beginners. Whether you want to grow vegetables, herbs, or flowers, I'll show you how. You don't need to be an expert or have a large garden to succeed. With practical, easy-to-follow advice, you'll be able to nurture your own container garden.

Container gardening offers many benefits. It's flexible and space-efficient. You can move your containers to catch the sun or bring them inside when it's cold. You have control over the growing conditions, which means healthier plants. Plus, there's nothing quite like picking fresh produce from your own garden or enjoying the scent of blooming flowers.

I understand the concerns many beginners have. You might worry about not having enough experience or space. Rest assured, this book addresses these challenges. I'll provide practical solutions to help you get started and succeed. You'll learn about choosing the right containers, caring for your plants, and using eco-friendly practices. We'll cover watering techniques, soil selection, and plant care. By the end, you'll have the knowledge and confidence to create your own garden.

Container gardening is not just a hobby; it's a journey of growth and experimentation. Embrace the process. There will be successes and failures, and both are valuable. Each step you take is a chance to learn and improve. Enjoy the journey and watch your efforts blossom.

I invite you to join me on this adventure. Let's transform your space into a thriving container garden. Whether you have a balcony, patio, or just a sunny windowsill, you can cultivate your own green space. Together, we'll explore the joys of gardening and discover the transformative power it holds. Welcome to the start of your container gardening journey.

CHAPTER 1

GETTING STARTED WITH CONTAINER GARDENING

UNDERSTANDING CONTAINER GARDENING: MORE THAN JUST A GARDEN IN A POT

Container gardening is not just about sticking plants in pots; it's a versatile method that adapts to your lifestyle. At its core, it involves growing plants in containers rather than directly in the ground. This technique allows you to cultivate a variety of plants, from leafy greens to bright blooms, without needing a sprawling yard. The beauty of container gardening lies in its flexibility, making it perfect for urban dwellers with limited space.

One of the primary benefits of container gardening is mobility. You can move your plants to catch the sun or bring them indoors when the weather turns nasty. This versatility means you can grow plants in places you might not have considered, like windowsills or rooftops. It's also incredibly accessible. You don't need a lot of tools or expertise to get started—just some containers, soil, and seeds. In urban environments, container gardening offers a solution to space constraints, letting you enjoy fresh produce and beautiful plants right at home.

However, like any endeavor, it comes with its challenges. Space-saving is a significant advantage since you can grow vertically or stack pots, but it also means you're working with limited soil volume. This can affect nutrient retention and water supply. Containers dry out faster than ground soil, so keeping your plants hydrated becomes crucial. But don't worry—these challenges are easily managed with the right strategies.

The creativity that container gardening allows is unmatched. You can arrange containers in various ways to suit your aesthetic preferences and functional needs. Consider mixing edibles like herbs and vegetables with ornamentals for a visually stunning and practical garden. The possibilities are endless—hanging baskets, tiered setups, or even repurposed items like teapots or boots can become unique planters. Your garden becomes an extension of your personality, allowing for endless experimentation.

Embracing sustainability is also key in container gardening. Upcycling materials for containers not only gives them a second life but also adds character to your garden. Look around your home for items that could be transformed into pots—a chipped mug or an old crate can work wonders. Sustainable practices extend to soil and fertilizer choices as well. Opt for organic options that nourish your plants while being kind to the environment.

Container gardening is both an art and a science, offering a rewarding experience for those willing to dive into its world. You'll find that each plant you nurture teaches you something new about growth and resilience. As you experiment and learn, remember that mistakes are part of the process. They lead to discoveries and solutions that improve your gardening skills.

The benefits you reap from container gardening go beyond the fresh herbs or bright flowers you grow. It's about creating a slice of nature that brings joy and tranquility to your life, no matter where you live. This chapter is just the beginning of your journey into this green adventure, where every pot holds the potential for something beautiful.

As we continue, we'll explore the practical aspects of setting up your container garden—choosing the right containers and soil, essential tools, and planning your layout. Each step will build your confidence and skills, turning your space into a thriving garden filled with life and color.

So let's get started on this exciting path, transforming your space one pot at a time into an oasis of beauty and growth.

CHOOSING THE RIGHT CONTAINERS: MATERIALS, SHAPES, AND SIZES

When it comes to container gardening, selecting the right containers is a pivotal decision that sets the foundation for your plant's health and growth. Let's start with materials. Terracotta pots are classic choices, known for their earthy aesthetic and breathability. They allow air and moisture to flow through the sides, which can help prevent root rot. However, they tend to be heavy and can crack in cold weather. Plastic containers, on the other hand, are lightweight and affordable. They retain moisture well, making them suitable for plants that require consistent hydration. But, they might not offer the same breathability as terracotta. Metal containers bring a modern touch to your garden with their sleek appearance, yet they can heat up quickly in direct sunlight, potentially harming sensitive roots. Understanding these nuances helps you decide what's best for your garden's needs.

Considering size and shape is just as crucial. The container's size dictates how much room your plant's roots have to grow. A container that is too small can stunt growth, while one that is too large can lead to waterlogging if the soil retains too much moisture. When choosing a container, ensure it has adequate drainage holes. Good drainage prevents water from sitting at the bottom, which can drown roots. For larger plants or those with extensive root systems, like tomatoes or peppers, opt for deeper pots. Shallow containers work well for herbs or succulents that don't require much root space. The shape of the container not only affects aesthetics but also how you arrange your plants. Tall, narrow pots can create height and drama, while wide, shallow ones can accommodate a lush arrangement of ground cover plants.

Beyond practicality, consider the aesthetic and functional aspects of your containers. Your garden should reflect your personal style and fit seamlessly into your living space. If your home leans towards a minimalist design, sleek metal or simple ceramic pots might complement it best. For a rustic feel, terracotta or wooden containers add warmth and texture. Space efficiency is also key, especially in small areas. Vertical planters or stacking pots can maximize space by allowing you to grow upwards rather than outwards. These options not only save space but also create eye-catching displays.

Maintaining your containers ensures they remain in top condition season after season. Regular cleaning prevents disease spread and keeps them looking fresh. For terracotta and ceramic pots, scrub them with a mixture of water and vinegar to remove mineral deposits and algae. Plastic containers benefit from a simple soap and water wash, while metal pots may require a gentle wipe to prevent rusting. Proper storage extends a container's lifespan; during harsh weather, store them in a shed or garage to protect them from cracking due to freezing temperatures.

Container Selection Checklist

- **Choose Materials Wisely**: Consider the pros and cons of terracotta, plastic, and metal.
- **Assess Size and Shape**: Ensure adequate root space and drainage.
- **Blend Aesthetics with Functionality**: Match designs with home decor and maximize space.

- **Regular Maintenance**: Clean thoroughly and store properly to extend lifespan.

Selecting containers thoughtfully sets your garden up for success. It's not just about where you place your plants but how you support their journey from seedling to mature beauty. With these considerations in mind, you're better equipped to create a thriving container garden that mirrors your personal style and meets your plants' needs, turning every pot into a statement of growth and creativity in your space.

CREATING THE PERFECT SOIL MIX FOR CONTAINERS

Soil is the foundation of any garden, but in container gardening, it plays an even more critical role. Unlike garden beds, where soil can naturally replenish nutrients and moisture, containers are limited environments. This means that the soil you choose must be top-notch. It needs to retain enough nutrients to feed your plants while having excellent drainage and aeration. Poor-quality soil can lead to waterlogged roots or nutrient deficiencies, which spell trouble for your plants. Think of your container soil as a blank canvas— what you mix into it determines the masterpiece you'll create.

A well-rounded soil mix starts with the basics: peat moss, perlite, and vermiculite. Peat moss serves as the backbone by retaining moisture and providing a lightweight structure. Perlite, those little white particles you often see, improves aeration and drainage, essential for healthy root growth. Vermiculite adds another layer of water retention but also aids in nutrient retention, keeping your plants well-fed. These three components work in harmony to create a balanced environment within your containers.

Beyond these basics, organic additives like compost enrich the soil with essential nutrients. Compost is nature's way of recycling nutrients back into the soil, providing a slow-release fertilizer that benefits your plants over time. It adds microorganisms that break down organic matter, boosting soil health. Incorporating compost into your mix not only enhances nutrient content but also improves moisture retention and aeration. For those seeking an extra nutrient boost, a sprinkle of worm castings can work wonders, adding beneficial bacteria and nutrients.

Crafting your own soil mix is both satisfying and cost-effective. Start with a basic recipe: two parts peat moss, one part perlite, one part vermiculite, and a generous portion of compost. Adjust these ratios based on your plant's specific needs; for instance, more perlite for succulents that require excellent drainage or more compost for nutrient-hungry vegetables. Mix thoroughly in a large container or wheelbarrow, ensuring an even distribution of materials. This DIY approach allows you to tailor the mix to your plants, optimizing their growth conditions.

Testing and amending your soil is crucial for maintaining optimal plant health. Begin by assessing the pH level, which determines nutrient availability. Most plants prefer a slightly acidic to neutral pH, around 6 to 7. Simple pH testing kits are available at garden centers and provide quick results. If adjustments are necessary, adding lime can raise the pH, while sulfur can lower it. These amendments should be made gradually and with caution, as drastic changes can shock your plants.

Soil Testing Tips

- **Use a Reliable Kit**: Invest in a quality pH testing kit for accurate readings.
- **Test Regularly**: Check pH levels seasonally to adapt to changes.
- **Amend Gradually**: Adjust pH levels slowly over time to prevent plant stress.

Regularly refreshing your soil is another key component of container gardening success. Over time, soil can become compacted and depleted of nutrients. Every season or two, consider emptying containers and mixing in fresh compost or new soil mix to invigorate the growing medium. This practice ensures that your plants have a continuous supply of nutrients and an environment conducive to healthy root development.

While creating the perfect soil mix may seem daunting at first, it's easier than it appears. Focus on understanding the needs of your plants and adjusting accordingly. A thoughtful approach to soil preparation sets the stage for vibrant growth and flourishing gardens.

The beauty of container gardening lies in these small yet impactful choices that empower you to cultivate thriving plants. Your attention to detail in selecting and mixing soil reflects your commitment to nurturing life in every pot and planter you tend.

By prioritizing soil quality and understanding its nuances, you're not just planting seeds but cultivating an environment where they can thrive and reach their full potential.

ESSENTIAL TOOLS FOR THE CONTAINER GARDENER

Stepping into the world of container gardening, it's easy to get overwhelmed by the array of tools available. However, you don't need a shed full of gadgets to succeed. Start with the basics, and you'll find that the right tools can make your gardening experience both enjoyable and productive. Let's explore a few must-haves that every beginner should consider adding to their collection.

Typical trowel

First on our list is the trusty trowel. This small but mighty tool will become your best friend as you dig into various tasks. From planting to transplanting and even weeding, a trowel's pointed blade allows you to navigate tight spaces with ease. Look for one with a comfortable handle that fits snugly in your hand, reducing strain during extended use. To keep your trowel in top condition, rinse it after each use to remove soil and debris, then dry it thoroughly to prevent rust. Store it in a dry place, preferably hanging on a peg or tucked away in a tool caddy.

Next up is the watering can. While any container can technically be used for watering, choosing a can with a long spout gives you precise control over where the water goes. This feature is particularly useful for reaching deep into pots without drenching the leaves, which can lead to disease. Opt for a can that holds enough water for your needs but isn't so heavy when full that it becomes cumbersome to carry. After each use, empty any remaining water to prevent algae buildup and store it upside down to ensure it dries completely.

Pruners are another essential tool in your gardening arsenal. These sharp scissors help you trim and shape plants, encouraging healthy growth and preventing disease spread. When selecting pruners, choose a pair that feels balanced and comfortable in your hand. For maintenance, clean the blades with soapy water after use and apply a light coat of oil to keep them lubricated and rust-free. Regular sharpening ensures they're always ready for action.

While these three tools form the core of your gardening kit, there are a few extras that can enhance efficiency and ease your workload. Self-watering systems, for instance, are a game-changer for those who travel frequently or have busy schedules. These systems deliver consistent moisture directly to the plant's roots, reducing the risk of overwatering and keeping your plants happy even when you're away.

A portable gardening bench is another handy addition. It provides a comfortable spot to sit or kneel while working on your plants, and many models include storage compartments for your tools. This setup keeps everything you need within arm's reach and saves you from unnecessary trips back and forth.

For those looking to streamline their tasks further, consider investing in quality gloves to protect your hands from thorns and blisters. A soil scoop can make transferring potting mix into containers less messy, while plant labels help you keep track of what's growing where. Each tool you add should serve a purpose and make your gardening experience more enjoyable.

. . .

Tool Maintenance Tips

- **Clean Regularly**: Rinse soil off tools after use.
- **Dry Thoroughly**: Prevent rust by drying tools completely.
- **Store Properly**: Hang or place tools in a dry area to avoid moisture damage.

Remember, it's not about having every tool on the market but rather selecting those that suit your specific needs and enhance your gardening experience. Over time, as you become more comfortable with container gardening, you might find yourself drawn to specialty tools that cater to unique tasks or particular plant types. Embrace this process of discovery—each tool you add is another step towards creating the garden you've envisioned.

Container gardening is about making the most of what you have, both in terms of space and resources. With these essential tools by your side, you'll be well-equipped to tackle any gardening challenge that comes your way. Whether you're planting seeds or nurturing mature plants, these tools will support your efforts and help ensure success.

PLANNING YOUR GARDEN LAYOUT: MAXIMIZING SMALL SPACES

In the realm of container gardening, optimizing space is a thrilling challenge, akin to piecing together an intricate puzzle. Picture your small balcony or patio as a blank canvas, a space brimming with potential just waiting to be unlocked by your creativity and careful planning. Vertical gardening becomes your paintbrush, allowing you to add both layers and depth to your design landscape. With a handful of the right techniques, you can transform this limited area into a lush, tranquil green sanctuary. Consider using sturdy materials like reclaimed wood for shelves or rustic iron for hanging baskets to elevate your plants, crafting a tiered effect that draws the eye upward and creates a sense of expansiveness. Walls and railings naturally serve as invaluable allies; they provide robust support for climbing plants such as sweet peas or morning glories, which can cascade gracefully over the edges, adding not just texture but also an undeniable charm and vibrancy to your space.

The arrangement of your plants in this miniature Eden is crucial, not only for aesthetics but also for achieving full functionality. Think of your garden as a living mosaic, a canvas of ever-shifting patterns and hues. Start by thoughtfully coordinating colors to create visual harmony that pleases not only the eye but also the soul. Group plants with similar hues—deep purples with lush greens or sunny yellows with vibrant reds—or choose contrasting shades for a striking, unforgettable effect. Height also plays a significant role; wisely place taller plants like verdant tomatoes or radiant sunflowers towards the back or center, with their shorter counterparts, such as fragrant herbs, in front. This creates a meticulously layered look, ensuring each plant, with its innate beauty, gets its moment in the spotlight without overshadowing its neighbors. Consider the changing seasons in your layout. Plan so that as one plant fades into dormancy, another comes into vibrant bloom, sustaining a dynamic and ever-evolving landscape.

Sunlight, the lifeblood of any thriving garden, holds the power to make all the difference in your gardening endeavor. Positioning your plants to maximize sunlight involves a bit more than simply placing them in direct sunlight. Consider the natural orientation of your space. A south-facing window or balcony typically receives the most sunlight throughout the day, creating the perfect environment for sun-loving plants like fiery peppers or aromatic basil. Meanwhile, east or west-facing spaces may catch only partial sun, which accommodates shade-tolerant plants such as feathery ferns or crisp lettuce. Remember, even within small areas, intriguing microclimates exist. Observe and understand how light shifts and varies in different spots, perhaps due to shading from neighboring trees, buildings, or even architectural features of your structure.

Microclimates are truly fascinating phenomena; they offer distinct growing conditions that, when explored wisely, can be leveraged to your advantage. Perhaps you discover a nook that stays warmer during the brisk winter or cooler amid the scorching summer heat. Use these subtle nuances to your benefit when deciding the perfect placements for your containers. For instance, a spot near a brick wall might retain heat longer into the evening, nurturing warmth-loving plants to flourish. Conversely, areas that naturally catch soothing breezes can help cool down heat-sensitive species during the hottest months of the year, ensuring their comfort and survival.

INTERACTIVE LAYOUT EXERCISE

Let's dive into a hands-on approach. Grab a piece of paper and sketch your space to embark on this creative journey. Plot where sunlight falls at different times of day, marking areas that bask in morning light against those shaded in the afternoon. Identify any existing structures like walls or railings, noting them as potential assets in your plan. Visualize how you might incorporate vertical elements, or how you could use color and height to design your living masterpiece. This strategic exercise primes you to plan thoughtfully and efficiently before purchasing plants, optimizing not only your space but also your budget.

By embracing these detailed techniques, your small space can transform into a thriving garden replete with layers of interest, beauty, and personality, all tailored to your unique environment and creative vision. In no time, your small yet mighty garden will stand as a testament to your dedication, thriving in flawless harmony with its surroundings.

SETTING REALISTIC EXPECTATIONS: WHAT TO GROW AND HOW MUCH

Stepping into the world of container gardening can be both exhilarating and daunting. It's essential to set realistic goals that honor your space, time, and resources. Start small. Imagine you're dipping your toes into the water rather than diving headfirst. Herbs like basil, mint, and parsley are perfect for beginners. They're forgiving, grow quickly, and offer immediate satisfaction as you snip fresh leaves for your culinary creations. If you're inclined towards vegetables, consider starting with lettuce or radishes. These are quick to harvest and don't demand much space. A small success here and there boosts confidence and sets the stage for more ambitious projects. (We'll delve more into selecting plants in the next chapter.)

When you begin this venture, understanding plant growth cycles is crucial. Each plant has its own rhythm and timeline. For example, tomatoes may take a few months to mature, while radishes can be ready in just a few weeks. Knowing these timelines helps you plan your garden activities and manage expectations. Additionally, be aware of the expected yield per container. A single tomato plant might fill a large pot and yield several pounds of fruit throughout the season, whereas a basil plant in a smaller pot could provide fresh leaves for months with regular pruning. Understanding these cycles allows you to plan meals around your harvest.

As you gain experience, scaling your garden becomes a natural progression. It's like building a collection of favorite books—one addition at a time. You might start with a few pots on the windowsill, then expand to hanging baskets or larger containers as you grow more confident. Introduce new plants gradually, allowing yourself time to learn their specific needs. Maybe add a pot of cherry tomatoes or a container of vibrant marigolds to attract pollinators. Slowly, your garden transforms into a thriving ecosystem.

Rotational planting is an excellent strategy as you expand. This technique involves swapping out plants as they finish their growth cycle to make the most of your space. For instance, after harvesting spring lettuce, replace it with summer peppers or eggplants. This approach keeps your garden productive throughout the year, offering a variety of crops and flowers to enjoy each season.

Growth Timeline Reference

To help you plan better, here's a simple chart of growth timelines for common container plants:

- **Basil**: 4-6 weeks until first harvest
- **Tomatoes**: 60-80 days until first harvest
- **Radishes**: 3-4 weeks
- **Lettuce**: 4-6 weeks
- **Peppers**: 70-85 days

This list isn't exhaustive, but it gives you an idea of what to expect as you start planting.

Finally, remember that gardening is as much about the experience as it is about the results. Every plant teaches you something new, whether it's how to adjust watering schedules or how sunlight shifts throughout the year. Celebrate every sprout and bloom as milestones in your gardening adventure. Mistakes will happen— they're part of the learning process. Embrace them as opportunities for growth rather than setbacks.

As you nurture your plants, you'll find that your garden becomes an extension of yourself—a reflection of your care and creativity. Whether you're growing herbs for the kitchen or flowers for the sheer joy they bring, every effort contributes to cultivating beauty in your life.

With these insights and strategies at hand, you're ready to set realistic expectations and grow your garden with confidence and enthusiasm. Each step you take enriches not just your space but also your understanding and appreciation of nature's wonders.

CHAPTER 2
SELECTING AND CARING FOR PLANTS

BEST VEGETABLES FOR CONTAINER GARDENING: A BEGINNER'S GUIDE

I magine waking up, stepping onto your small balcony, and picking fresh lettuce leaves for your lunchtime sandwich that day. This isn't just a reverie confined to dreams—it's something utterly tangible and achievable, even if you're stepping into the world of gardening without any prior experience. Container gardening unveils a realm of endless possibilities, especially when it comes to cultivating your very own vegetables. Let us embark on a journey to explore a selection of vegetables that are exceptionally easy to grow, making them perfect candidates for beginners like you. Leafy greens such as lettuce and spinach emerge as fantastic choices. They thrive admirably within the confines of a container and exhibit swift growth, gracing you with a continuous and plentiful supply of fresh, verdant leaves. These plants exhibit a remarkable resilience and adaptability; they don't demand much space, rendering them ideal for urban gardening settings where space is often a luxury. In addition to their minimalistic space needs, they're forgiving companions who won't hold a grudge if you happen to miss a watering or two.

Root vegetables, including radishes and carrots, have also proven their mettle in container settings. Radishes, with their rapid growth cycle, are tailor-made for the impatient gardener who eagerly anticipates visible progress. Carrots, while calling for deeper containers to accommodate their length, reward your patience with crunchy textures and a burst of sweet flavor. Opting for these particular vegetables is a wise starting point for embarking on your container gardening journey. The key to success lies in selecting compact, adaptable varieties that harmonize with limited spaces. When browsing seed catalogs or garden centers, keep an eye out for varieties labeled as "baby" or "dwarf," as they are expertly tailored for small spaces and will still yield a gratifying harvest.

Understanding the intricate space and light requirements of your chosen vegetables stands as a cornerstone of successful container gardening. Compact living spaces necessitate meticulous planning, urging you to select varieties that align with your spatial limitations. If you happen upon a sunny corner or spot on your patio, seize the opportunity by planting sun-loving vegetables like tomatoes or peppers that bask and grow luxuriously under ample sunlight. Conversely, if your area leans toward the shaded side, lean towards leafy greens that exhibit a remarkable tolerance for lower light levels and continue to flourish.

Growing vegetables in containers is a straightforward process that begins with selecting the appropriate container size. This initial step is crucial as it ensures your seeds have the necessary space and depth to thrive. For beginners, understanding that different vegetables have varying requirements is key. For instance, leafy greens such as lettuce and spinach are particularly adaptable and can be directly sown into pots without much fuss. These vegetables don't require deep soil to flourish, making them ideal choices for your first foray into container gardening. When planting radishes, it's important to space them a few inches apart. This spacing allows each radish enough room to grow both above and below the soil. Crowded plants often fail to develop properly, leading to a disappointing harvest. Therefore, paying attention to the spacing recommendations for each type of vegetable you plant can significantly impact your gardening success. Carrots, on the other hand, demand pots that are deeper than what most other vegetables require. This is because carrots need to extend their roots deep into the soil as they grow. Choosing a container that is too shallow will stunt their growth and result in undersized carrots. It's advisable to opt for pots that are at least 12 inches deep for these root vegetables, ensuring they have ample space to reach their full potential. Watering your container garden consistently is another critical aspect of growing vegetables. Uneven watering can lead to various problems, including bolting, where plants grow quickly but produce fewer leaves and flowers, or the development of bitter-tasting greens. To avoid these issues, establish a regular watering schedule. Vegetables in containers often require more frequent watering than those in the ground, as soil in pots can dry out quickly, especially in warm weather. Monitor the soil moisture level closely, and adjust your watering routine as needed to keep the soil evenly moist, not waterlogged or bone dry. By following these guidelines, you'll set the stage for a productive and enjoyable container gardening experience.

As you delve into container gardening, prepare for potential hurdles like pests or bolting. Aphids may become unwelcome tenants on your plants, but there's no need for alarm—natural remedies stand at the ready to address these nuisances. Employing neem oil or concocting a homemade soap spray stands as an effective measure to keep them at bay, steering clear of harsh chemicals and aligning with sustainable gardening practices. Bolting, mentioned in the last paragraph, also tends to occur in warmer weather. To counteract this, consider scheduling the planting of cool-season crops either early in spring or later in the fall when milder temperatures prevail. (We'll delve into common pests and diseases more in Chapter 5.)

Closeup of an aphid

REFLECTION EXERCISE: YOUR IDEAL VEGETABLE GARDEN

Pause for a moment to conjure an image of your ideal vegetable container garden. Consider what delights you envision cultivating, and meticulously jot down the vegetables that ignite your excitement the most. Take into account their space and light prerequisites for an informed and strategic plan. Reflect on the harmony between these chosen vegetables and the space available to you, contemplating any necessary adjustments to ensure their optimal growth.

By embarking with these beginner-friendly vegetables, you are meticulously laying the foundation for a flourishing venture into container gardening. The unparalleled joy of harvesting produce nurtured by your own hands and care is one of life's simple pleasures. With a sprinkle of dedication, a dash of attention, and an abundance of enthusiasm, you'll soon revel in the magnificent fruits (and vegetables) of your labor, savoring a vibrant, homegrown bounty.

HERBS THAT THRIVE IN CONTAINERS AND SMALL SPACES

Imagine a sunny afternoon, and you're in your kitchen preparing a meal. You reach over to your windowsill and pluck a few fresh basil leaves, their aroma instantly filling the air. This is the charm of growing herbs in containers, making them accessible and practical for any meal. Herbs like basil, mint, and parsley are not only easy to grow but also become integral to your culinary repertoire. Basil, with its lush, aromatic leaves, thrives in warm conditions and adds a sweet, peppery flavor to dishes. Mint spreads quickly, offering a refreshing taste perfect for teas or garnishes. Parsley, appreciated for its versatility, complements a wide range of recipes. These herbs are beginner-friendly, forgiving of slight neglect, and bounce back with a bit of care.

For those seeking more lasting companions, consider perennial herbs like thyme and rosemary. These resilient plants add depth to your garden with their woody stems and fragrant leaves. Thyme's small, hardy leaves pack a punch of flavor, while rosemary's needle-like foliage releases a rich aroma, perfect for roasts or stews. Both herbs thrive in well-drained soil and sunny spots, rewarding you with year-round harvests without demanding constant attention.

Planting and caring for herbs require understanding their unique needs. Good drainage is paramount; ensure your containers have holes to prevent waterlogging. Most herbs prefer full sun but can tolerate partial shade. The key to thriving herbs lies in regular harvesting. Snipping leaves not only provides you with fresh ingredients but also encourages new growth, keeping plants bushy and productive. Be mindful of overwatering—herbs prefer slightly dry conditions between watering sessions.

The choice between indoor and outdoor growing depends on your space and preferences. Indoor herb gardens thrive on sunny windowsills, offering convenience and quick access while cooking. However, they may need supplemental lighting during darker months. Outdoor options like balconies or patios provide more room for expansive growth, allowing herbs to bask in natural sunlight and fresh air. Outdoor settings also offer the opportunity to companion plant with other container vegetables or flowers.

Propagation is a rewarding way to expand your herb collection without breaking the bank. Mint and basil are particularly easy to propagate using cuttings. For mint, select a healthy stem around 4 inches long, remove the lower leaves, and place it in water until roots form. Once roots are visible, transfer it to a pot with well-draining soil. Basil follows a similar process; simply snip a stem below a leaf node, and watch as roots develop in water. These techniques not only extend your garden but also offer endless opportunities for sharing plants with friends or experimenting with new varieties.

Growing herbs in containers is a delightful endeavor that brings fresh flavors and vibrant scents to your home. It's about more than just planting seeds; it's creating a living pantry that enhances your meals and adds beauty to your space. Whether you're an avid cook or simply enjoy the occasional fresh garnish, container gardening allows you to cultivate an array of herbs tailored to your tastes and lifestyle.

FLOWERING PLANTS FOR YEAR-ROUND COLOR AND INTEREST

Marigolds, with their fiery orange and yellow petals, thrive in containers and are ideal for beginners. They're hardy, resist pests, and bloom throughout the season. Petunias add a touch of elegance with their continuous blooms, painting your garden in shades of pink, purple, and white. These flowers don't just look pretty; they bring your space to life, inviting pollinators like bees and butterflies.

To keep your garden lively all year, consider planting with the seasons in mind. Cool-season flowers like pansies and violas flourish in the fall and winter. They withstand cooler temperatures and add much-needed color during the dreary months. As spring warms up, switch to warm-season blooms like zinnias and cosmos. These plants bask in the sunlight and thrive in the heat, ensuring your garden stays colorful through summer. By rotating flowers with the seasons, you maintain a dynamic garden that never loses its charm.

Arranging flowers for visual impact is an art form. Mixing colors and textures can transform your space into a living master-piece. Combine bold hues with softer pastels for a striking contrast. Consider the texture of leaves and petals—pairing smooth with jagged edges creates an intriguing display. Trailing plants like sweet potato vine or ivy add height and depth, cascading over the edges of containers and drawing the eye. Layering plants by height ensures each one gets its moment to shine, creating a balanced and harmonious arrangement.

Flowering plants come with their own set of challenges, but with a little know-how, you can tackle them head-on. Deadheading, or removing spent blooms, encourages continuous flowering. This simple task redirects energy from seed production back into growth, keeping your plants vibrant. Pests like aphids might make an appearance, but don't fret. Natural solutions like neem oil or insecticidal soap keep them under control without harming beneficial insects. Regularly inspect your plants for signs of trouble—a quick intervention can prevent bigger issues down the line.

Whether you're new to gardening or a seasoned pro, flowering plants offer endless opportunities to experiment and express yourself. They bring joy not just through their blooms but also through the process of nurturing them to life. As you explore different varieties and arrangements, remember that gardening is as much about the journey as it is about the destination. Each season brings new challenges and rewards, keeping your passion for gardening alive.

Creating a garden filled with year-round color isn't just about choosing the right plants; it's about understanding how they interact with each other and their environment. Light, water, and soil all play critical roles in plant health. Pay attention to these elements, adjusting as needed to suit each plant's preferences. For instance, marigolds thrive in full sun while pansies prefer partial shade. Matching plants to their ideal conditions ensures they put on their best show.

Your garden reflects your personality—a canvas where you can play with colors, textures, and forms. Don't be afraid to experiment with unconventional combinations or try new plants each season. The beauty of container gardening lies in its flexibility; you can rearrange plants or swap them out as your tastes change. This adaptability keeps your garden fresh and exciting, a true reflection of your evolving style.

As you explore the world of flowering plants, you'll find that each one has its own story to tell. From towering sunflowers to delicate violas, they bring diverse colors and shapes to your garden. Embrace their uniqueness, learning from their successes and setbacks alike. With patience and care, you'll cultivate not just a garden but a living testament to your creativity and passion for nature's beauty.

COMPANION PLANTING IN CONTAINERS: PLANTS THAT GROW WELL TOGETHER

Companion planting is an ancient gardening practice that can transform your container garden into a thriving ecosystem. Imagine your plants as friendly neighbors who help each other grow better and stay healthy. By placing them together, you can naturally repel pests and enhance flavor, all while maximizing space. Picture tomatoes and basil growing side by side. The basil has a knack for deterring pesky insects that often target tomatoes. In return, the tomatoes provide a bit of shade, helping basil thrive. Plus, the flavor combination is unbeatable when it comes to making sauces or salads.

Let's consider carrots and onions. This duo works like a charm. Onions emit a scent that can confuse and deter carrot root flies, keeping your carrots safe from pests. These two not only protect each other but also make efficient use of space since they grow at different levels—onions above ground and carrots below. This is what companion planting is all about: creating harmonious partnerships that benefit everyone involved.

When planning your garden, it's vital to think about the available space and how plants interact underground. Roots need room to spread and access nutrients without competing aggressively. Layering plants based on root depth can help manage this effectively. For instance, shallow-rooted herbs can share a pot with deep-rooted vegetables because they won't compete for the same resources. Always ensure each plant has enough room for its roots to expand comfortably.

Experimentation is the spice of gardening. Don't hesitate to try new combinations and see what works best in your containers.

Keep a gardening journal to track what you plant together, how they perform, and any changes you notice. Write down everything from the date of planting to any pests you spot. Over time, you'll gather a wealth of knowledge about what specific pairings thrive in your unique conditions.

This approach not only makes gardening more engaging but also helps you learn from both successes and mishaps. You might find that some plants unexpectedly flourish together, while others don't get along as well as anticipated. The beauty of container gardening is its flexibility—if something doesn't work, it's easy to rearrange plants or try different groups without much hassle.

Beyond the practical benefits, companion planting adds an element of creativity to your garden design. Mix different colors and textures to create visually stunning displays. Let trailing plants like nasturtiums spill over the edges of pots while tall sunflowers stand proudly behind them. This layering not only maximizes space but also enhances the aesthetic appeal of your garden.

As you delve deeper into companion planting, you'll discover that certain combinations offer surprising perks. For instance, marigolds planted near cucumbers can deter nematodes and attract pollinators, boosting overall plant health and yield. Similarly, lettuce planted alongside chives benefits from the chives' natural pest-repellent properties while enjoying the shade provided by taller plants.

One thing to remember is that not all plant pairings are beneficial. Some combinations can lead to increased competition for resources or attract pests rather than deter them. Avoid pairing dill with carrots, as they compete for nutrients and may attract undesirable insects.

The key to successful companion planting is understanding the needs and behaviors of each plant species. Invest time in observing how they interact with one another and their environment. Pay attention to factors like sunlight exposure, watering schedules, and soil conditions.

Experimenting with companion planting not only enhances your garden's productivity but also deepens your understanding of plant interactions and ecosystems. As you continue exploring different combinations, you'll develop a keen sense of what works best in your specific environment, allowing you to create a thriving container garden that's both beautiful and productive.

SEASONAL PLANTING: TIMING YOUR PLANTING FOR OPTIMAL GROWTH

Timing is everything in gardening. Just like planning a vacation or scheduling your favorite show, knowing when to plant can make or break your success. Seasonal planting is about aligning your gardening efforts with nature's rhythms. It's about understanding frost dates and growing seasons, those pivotal moments that dictate when seedlings can safely venture outside. Every region has its own climate quirks, and getting familiar with yours is like learning the secret handshake to successful gardening.

Start by marking frost dates on your calendar. The last frost date is typically when it's safe to plant tender seedlings outdoors, while the first frost date signals the end of the growing season. These dates vary based on your location, so a little research goes a long way. Online tools and gardening apps make it super easy to find this information, often providing regional planting guides tailored to your area. Armed with this knowledge, you can plan your planting calendar like a seasoned pro.

Creating a planting schedule doesn't have to be as complicated as it sounds. Begin by jotting down what you want to grow and when it should be planted. Consider the time each plant needs to mature and work backward from the first frost date to determine the best planting time. This approach allows for rotational planting, where you stagger plantings of short-season crops like radishes or lettuce. As one batch finishes, sow another, ensuring a continuous harvest. This technique maximizes yield and keeps your garden productive throughout the year.

Succession planting is another gem in the gardener's toolkit. It's all about keeping your garden buzzing with activity by planting new crops as others fade out. Imagine harvesting peas in early summer, then using the same space for bush beans or late-season greens. By staggering planting times, you create a dynamic garden that provides fresh produce over an extended period. This approach not only enhances productivity but also allows for experimenting with different crops and varieties.

Of course, nature has a mind of its own, and seasonal changes can throw a wrench in even the best-laid plans. Unexpected weather shifts are part and parcel of gardening, but with a little preparation, you can navigate these challenges with ease. Protecting plants from sudden cold snaps or heat waves involves creative solutions like row covers or shade cloths. These simple tools act as barriers against extreme conditions, providing a buffer that keeps plants happy and healthy.

Transitioning plants between seasons requires thoughtful attention. As temperatures drop in autumn, consider moving tender perennials into more sheltered spots or indoors if possible. Adjust watering schedules and reduce feeding as growth slows during cooler months. Preparing for seasonal transitions ensures plants remain robust and resilient, ready to flourish when conditions improve.

Gardening is a dance with nature, where timing and adaptability are your best allies. By tuning into seasonal rhythms and planning accordingly, you set the stage for a thriving garden that mirrors nature's cycles. Whether it's synchronizing with frost dates or embracing succession planting, these strategies empower you to make informed decisions that elevate your gardening game.

As you continue exploring seasonal planting, remember that each year brings new opportunities to refine your approach. Keep track of what works well and where improvements can be made.

This ongoing process of learning and adapting enriches your understanding of plant behavior and enhances your connection to the natural world.

In this gardening dance, every plant has its season—embracing this idea helps you cultivate a garden that's not only productive but also deeply rewarding.

PLANT CARE BASICS: PRUNING, DEADHEADING, AND BEYOND

Imagine your garden as a living tapestry, where each plant plays a vital role. Just like any masterpiece, it requires regular upkeep to maintain its vibrancy. Pruning and deadheading are two key techniques that help keep your plants looking their best. Pruning is akin to giving your plant a haircut—it shapes and maintains its health by removing dead or overgrown branches. This not only enhances airflow but also reduces the risk of disease. For most plants, you'll want to prune in early spring before new growth begins, using sharp pruners or scissors to make clean cuts just above a leaf node or bud.

Deadheading, on the other hand, focuses on flowers. It involves snipping off spent blooms to encourage new ones. This simple practice redirects the plant's energy from seed production back into flowering, prolonging the bloom period. Using a pair of small scissors or your fingers, pinch off the faded flowers just below the base. Deadheading can be done throughout the growing season, and it's a great way to keep your garden lively and colorful.

Proper tool usage is crucial for effective plant care. Invest in a good pair of pruners with ergonomic handles that fit comfortably in your hand. Keep them clean and sharp to ensure precise cuts that heal quickly. Timing is everything; prune shrubs and perennials in early spring, while flowering plants benefit from regular deadheading throughout their bloom cycle. Remember, each snip is a step toward a healthier garden.

Plants, much like people, exhibit signs of stress when something isn't right. Wilting leaves might suggest underwatering, while yellowing could indicate nutrient deficiencies or root problems. It's essential to observe these signals and respond promptly. For wilting, check soil moisture and adjust your watering schedule accordingly. If leaves yellow, consider adding a balanced fertilizer to replenish lost nutrients. Compost or organic fertilizers work wonders here, providing the essential elements plants need to thrive.

Routine observation plays a pivotal role in maintaining plant health. A quick daily check allows you to spot potential issues before they escalate. Inspect leaves for signs of pests like aphids or spider mites. These tiny invaders can wreak havoc if left unchecked, but a simple spray of water or insecticidal soap often resolves the issue. Regular monitoring also helps you notice positive changes—new growth, budding flowers, or vibrant colors—which is incredibly rewarding.

The benefits of these practices extend beyond aesthetics; they contribute to the overall well-being of your garden. Healthy plants are more resilient to pests and diseases, and they produce more abundant blooms and fruits. By dedicating a few moments each day to plant care, you cultivate a garden that not only looks beautiful but also thrives with life and energy.

As you hone your skills in pruning and deadheading, remember that gardening is an evolving process. Each plant has its own rhythm and preferences, so stay curious and attentive to their needs. Don't hesitate to experiment with different techniques or schedules—what works for one plant might not suit another.

In this chapter, we've explored essential plant care techniques that form the foundation of a flourishing garden. From shaping with pruning to encouraging blooms through deadheading, these practices empower you to nurture healthy plants that enrich your space. As you continue your gardening journey, keep these insights in mind and embrace the joy of tending to your green companions.

In our next chapter, we'll delve into watering techniques and strategies for maintaining optimal soil moisture—a crucial aspect of container gardening success. Until then, enjoy the time spent with your plants and the small victories that come with each new leaf or flower.

CHAPTER 3
WATERING AND FERTILIZING ESSENTIALS

MASTERING THE ART OF WATERING: TECHNIQUES FOR BUSY SCHEDULES

P icture this: you're juggling work assignments, managing family commitments, and carving out some precious me-time, yet you persist in nurturing a dream of a flourishing container garden. The palpable challenge remains: how do you ensure your plants stay happy and healthy without dedicating copious amounts of time each day? The secret lies in adopting efficient and effective watering techniques specifically tailored for those with bustling and hectic lifestyles, without sacrificing the vitality of your garden. One remarkably effective method is the utilization of a drip irrigation system.

A drip irrigation setup delivers water directly to the base of each plant through a network of tubing and emitters, minimizing waste and ensuring consistent moisture. A small pump, often placed inside a water reservoir, pushes water through connected tubing up to the drip line, which runs above or alongside the containers. Emitters positioned above each pot release water slowly, providing efficient hydration directly to the roots. The system can be automated with a timer.

This savvy setup automates the entire watering process, which could be revolutionary for anyone who struggles to keep up with daily watering demands, providing the freedom to attend to life's myriad other obligations with ease. With essential components like emitters and tubing, this method allows you to customize the flow to meet the unique needs of each plant, ensuring optimal hydration even when you are not physically present to tend to them.

In addition to technological solutions, employing nature-inspired strategies can also enhance your watering regimen. For example, another invaluable strategy includes the introduction of moisture-retaining mulch. This cost-effective and straightforward solution greatly aids in helping the soil retain water, thereby reducing the frequency and intensity of watering required. Mulch acts like a protective shield or blanket, covering the soil surface to dramatically minimize evaporation. Options such as straw, bark chips, or even coconut coir not only trap moisture but also contribute additional nutrients to the soil as they decompose, thereby enriching the overall soil quality over time. By simply spreading a generous layer around your plants, you'll observe that they remain moist for prolonged periods, even during the most oppressive heat spells.

Commitment to consistent watering schedules is yet another cornerstone of sound plant health management. Establishing regular reminders or leveraging technology through apps can aid significantly in maintaining regularity. By doing so, you success-fully mitigate the risks associated with both overwatering and underwatering. Many contemporary gardening apps provide features such as local weather forecasts and personalized alerts, keeping you informed about the best times to water your garden. Installing sensors to monitor soil moisture levels supplies addi-tional real-time insight and precision, as these compact devices measure the soil's moisture content and alert you when it's time for another round of watering. They eliminate guesswork from the equation, permitting you to center your efforts on simply enjoying the vibrant oasis that is your green space.

Hot weather scenarios introduce unique challenges, but with some foresight and preparation, you can masterfully keep your plants hydrated even through the most torrid heat waves. Engaging in watering sessions early in the morning or late in the afternoon effectively minimizes evaporation, ensuring the roots can absorb maximum moisture for sustained nourishment.

Recognizing and interpreting the signs of water stress is critical for promptly troubleshooting any potential issues that may arise. Wilting leaves often indicate inadequate water supply, but they may also be symptomatic of root complications if the plant does not recover post-watering. Similarly, yellowing leaves may suggest overwatering, nutrient deficiencies, or other concerns, indicating that an adjustment in your approach may be necessary. Taking meticulous note of these symptoms allows for quick, responsive action to effectively prevent any long-term detriment.

Understanding these indicators and intricacies empowers you to make well-informed, confident decisions regarding your garden's needs. By refining your watering techniques, embracing technology while honoring natural methods, and being astutely attuned to signs of stress, you'll nurture a robust and thriving container garden that seamlessly integrates into your busy lifestyle. Gardening is truly about harmonizing different aspects of life— between work obligations and leisure activities, between attentive care and negligence. Equipped with these strategies and insights, you're poised to cultivate a vibrant garden that not only flourishes but stands as a testament to your dedication, adaptability, and passion for finding beauty amidst the chaos of life. Each plant speaks volumes of your steadfast commitment, thriving under the attentive care you provide.

SELF-WATERING CONTAINERS: A TIME-SAVING SOLUTION

Self-watering containers reduce the frequency of watering, allowing you more time to enjoy the garden rather than maintaining it. By ensuring consistent moisture levels, these containers create a stable environment for plant roots, minimizing stress. This consistency means that even when life gets hectic, your plants stay happy and hydrated. These containers are a game-changer, especially for those who want a lush garden without the commitment of daily upkeep.

Choosing the right self-watering system is paramount to achieving these benefits. Several types are available, each with its own strengths. Wick-based systems use a simple fabric wick that draws water from a reservoir at the bottom of the container into the soil above. This method is effective for smaller plants and herbs that require steady moisture. Reservoir-based containers feature a built-in water reservoir beneath the soil layer. Water seeps upward through capillary action, providing uniform moisture distribution. Capillary action systems work similarly but use a specialized mat or layer that allows water to rise from below, keeping the soil evenly moist. Understanding these options helps you select a system that aligns with your garden's needs and your lifestyle.

Setting up a self-watering container may initially seem complex, but it's quite straightforward with a little guidance. Start by assembling the container according to the instructions provided, ensuring each component fits securely. Most models require positioning a reservoir at the base with an overflow hole to prevent excess water buildup. Fill this reservoir with water and place the soil on top, embedding the wick or mat to draw moisture upward. The key here is ensuring that the wick maintains contact with both water and soil for effective functioning. Regular maintenance involves checking and refilling the reservoir as needed, typically every couple of weeks. Cleaning is equally essential to prevent algae growth, which can clog systems and impede water flow. Empty the reservoir periodically and scrub it with mild soap, rinsing thoroughly before refilling.

While self-watering containers offer numerous advantages, they aren't without their quirks. Algae growth in reservoirs is a common issue, but can be managed by keeping the reservoir covered or using opaque materials to limit sunlight exposure, which fuels algae proliferation. Proper drainage is another crucial aspect; without it, plants risk becoming waterlogged, leading to root rot. Ensure excess water can escape through designated drainage holes and consider adding a layer of coarse material like gravel at the bottom to facilitate drainage. Address these concerns proactively, and self-watering systems will provide reliable support for your garden's hydration needs.

The beauty of self-watering containers lies in their capacity to adapt to various gardening situations and plant types. Whether you're cultivating herbs on a sunny windowsill or growing tomatoes on the patio, these containers simplify care routines while promoting robust plant development. You'll experience fewer watering mishaps, as plants receive just the right amount of moisture, preventing both underwatering and overwatering. As you integrate self-watering systems into your gardening practice, you'll discover newfound freedom and flexibility in how you manage your green space.

Embracing self-watering technology gives you the confidence to explore more diverse plant species without being tethered to a rigid watering schedule. It opens doors to experimenting with delicate plants that demand constant moisture or expanding your garden during peak summer months when maintaining hydration becomes challenging. These systems act as silent partners in your gardening endeavors—reliable allies that ensure your plants flourish while you focus on other priorities.

Incorporating self-watering containers into your gardening toolkit offers more than just convenience; it transforms how you interact with your plants, fostering an environment where they can thrive independently of constant attention. With these systems in place, you'll find yourself spending less time troubleshooting watering issues and more time enjoying the verdant beauty that surrounds you.

CHOOSING THE RIGHT FERTILIZER: ORGANIC VS. SYNTHETIC

In the world of container gardening, choosing the right fertilizer is like picking the right fuel for your car. It directly impacts how well your plants grow and thrive. Let's delve into the differences between organic and synthetic fertilizers, so you can make informed choices. Organic fertilizers, derived from natural sources like compost and manure, release nutrients slowly over time. This gradual release feeds plants steadily, reducing the risk of nutrient leaching and promoting sustainable growth. On the other hand, synthetic fertilizers offer a quick nutrient boost, ideal for plants needing immediate attention. They are manufactured from minerals and chemicals, providing a precise nutrient balance that acts fast. However, this rapid release can sometimes lead to nutrient runoff, impacting the environment negatively. The choice between slow and fast nutrient release depends on your plants' needs and how quickly you want to see results.

Environmental impact is another factor to consider. Organic fertilizers enhance soil health by stimulating beneficial microorganisms and improving soil structure. They contribute to long-term soil fertility and reduce the risk of chemical buildup. Synthetic fertilizers, while effective, can leave residues that might alter soil chemistry over time and potentially harm local waterways through runoff. For eco-conscious gardeners, organic options often align better with sustainable practices.

Selecting the right fertilizer involves matching nutrients to plant needs. For leafy greens like lettuce and spinach, high-nitrogen fertilizers are ideal as they promote lush foliage growth. Look for organic options like fish emulsion or blood meal for a steady nitrogen supply. When it comes to flowering plants or those producing fruits, phosphorus-rich fertilizers work wonders. They support robust root development and enhance bloom quality. Bone meal or rock phosphate are excellent organic choices, providing a gentle phosphorus boost without overwhelming the plant.

Application methods play a crucial role in fertilizer effectiveness. Granular fertilizers are easy to apply and release nutrients slowly as you water your plants. They are great for long-term feeding but require even distribution across the soil surface. Liquid fertilizers offer immediate results, ideal for plants needing a quick nutrient uptake. They are easy to mix with water and apply during regular watering sessions. Foliar feeding is another technique where you spray diluted fertilizer directly onto leaves for rapid absorption. This method is especially useful if you notice signs of nutrient deficiencies like yellowing leaves or stunted growth.

Timing is everything when applying fertilizers. Early morning or late afternoon applications help avoid nutrient evaporation in hot weather. For most container plants, a bi-weekly schedule works well, but adjust based on plant response and growth stage. Young seedlings may need less frequent feeding compared to mature plants actively producing fruits or flowers.

Avoiding over-fertilization is key to maintaining plant health. Excess nutrients can lead to leaf burn—a condition where leaf edges turn brown and crispy due to salt buildup from fertilizers. This stress not only affects aesthetics but also reduces overall plant vigor. Soil testing is a valuable tool for preventing over-fertilization. Simple kits available at garden centers provide insights into nutrient levels, helping you tailor fertilizer applications accordingly.

Understanding these nuances in fertilizer choice and application empowers you to nurture healthy, vibrant container gardens that flourish season after season. Balancing between organic and synthetic options allows you to customize care based on your gardening goals and environmental considerations. As you experiment with different fertilizers, observe how your plants respond and adjust your approach as needed, creating a harmonious relationship between nutrients and plant growth.

By embracing this approach, you'll develop an intuitive sense of what your garden needs at different stages, leading to thriving plants that reflect your care and attention to detail. With each application, you're not just feeding plants; you're fostering a lively ecosystem that rewards your efforts with beauty and bounty.

FEEDING YOUR PLANTS: NUTRIENT NEEDS AT DIFFERENT GROWTH STAGES

Understanding the nutrient needs of your plants as they grow is like learning the language of your garden. From the moment a seedling breaks through the soil to the time it matures into a robust plant, its nutritional requirements shift. Initially, as roots begin to develop, a higher phosphorus content is crucial. Phosphorus strengthens root systems, setting a solid foundation for the plant to thrive. As plants grow, their needs change. When it's time for fruit and flower production, potassium takes center stage. Potassium supports the development of blooms and fruits, ensuring they are healthy and abundant. This nutrient balance is critical throughout the plant's life cycle.

Creating a feeding schedule tailored to your plants is essential for keeping them healthy and productive. Fast-growing vegetables like tomatoes or peppers benefit from weekly feeding. This consistent nutrient supply supports their rapid growth and high energy demands. In contrast, slower-growing herbs such as thyme or rosemary do well with monthly feedings. These plants require less frequent nutrition boosts, allowing them to develop at their own pace without becoming overwhelmed by excess nutrients. Adjusting your feeding routine based on plant type and growth rate ensures that each plant receives the right amount of care.

Micronutrients, though needed in smaller quantities, play a significant role in plant health. Iron, for instance, is vital for chlorophyll production, the green pigment responsible for photosynthesis. Without iron, leaves may yellow, hindering the plant's ability to harness energy from sunlight. Magnesium also plays an essential role in photosynthesis by activating enzymes that facilitate energy transfer. Though these trace elements might seem minor, they are crucial for overall plant vitality. Ensuring your plants receive these micronutrients can make all the difference in their growth and resilience.

Sometimes, despite your best efforts, plants may show signs of nutrient deficiencies or excesses. Recognizing these signals is key to addressing issues before they escalate. Yellow leaves often indicate an iron deficiency. A simple iron supplement can correct this, restoring the plant's lush green color and vigor. Similarly, stunted growth may suggest a need for balanced fertilizers that provide a spectrum of nutrients. By being observant and responsive to these signs, you can tweak nutrient levels to suit your plants' needs better.

REFLECTION SECTION: TRACKING PLANT HEALTH

Keeping a garden journal can be invaluable in monitoring your plants' nutrient needs over time. Jot down observations about leaf color, growth rates, and any changes you notice after feeding. Note any adjustments made to your fertilizing routine and their outcomes. This practice not only helps track progress but also builds a wealth of personalized knowledge about your garden's unique requirements.

By embracing this understanding of nutrient needs throughout different growth stages, you empower yourself to cultivate a thriving container garden that flourishes season after season.

Balancing macro and micronutrients while adjusting feeding schedules based on plant health creates an environment where your plants can truly prosper.

Incorporating these insights into your gardening practice enriches your connection with nature and deepens your appreciation for the intricate dance of growth and nourishment that sustains life in your containers.

WATER CONSERVATION STRATEGIES FOR ECO-CONSCIOUS GARDENERS

Imagine walking into your garden, surrounded by lush greens and vibrant blooms, knowing with immense satisfaction that you're playing an active role in preserving our precious planet's resources. Water conservation is not merely a fashionable catchphrase or transient trend; rather, it's an essential practice imbued with profound significance, particularly for those among us with an acute awareness of our environmental impact. By employing simple yet effective strategies, we can make a significant difference.

One remarkably efficient method of conserving water involves the use of rainwater barrels. This environmentally-friendly and budget-conscious approach is as practical as it is ingenious. Visualize, if you will, strategically placing these barrels beneath your home's downspouts, where they discreetly yet effectively collect rainwater. During the dry spells, this reserve of rainwater stands ready to invigorate your garden, offering life-giving hydration without dipping into precious tap water reserves. Not only is this practice conducive to water saving, but it also furnishes your garden with natural, untreated water that is rich in nutrients and often more beneficial for plant health compared to treated alternatives.

In your pursuit of water conservation, another highly beneficial approach is utilizing greywater systems. These ingenious systems allow for the recycling of water from household sources such as baths, sinks, and washing machines, repurposing it for irrigation needs within your garden. By installing a straightforward grey-water system, you open the door to reusing household water, which serves to cut down on water waste and optimize available resources. Engaging in such a project can be particularly rewarding if you have a penchant for DIY solutions. Bear in mind the wisdom of utilizing eco-friendly soaps and detergents, ensuring that this recycled water remains a boon to your plants, rather than a detriment.

When curating your garden's collection of flora, it is wise to take into consideration plants that flourish with minimal water. A perfect inclusion to your environmentally mindful garden would be succulents and cacti, known for their natural ability to withstand arid climates. Their thick, fleshy leaves act as reservoirs, storing water and enabling survival even under harsh, dry conditions. Additionally, consider the inclusion of native plants—species that have adapted perfectly to your local climate and require minimal additional watering. These plants, steeped in the wisdom of adaptation, effortlessly navigate the vacillations of your region's weather patterns. In doing so, they not only bolster your water conservation efforts but also inject an array of vivid textures and hues into your garden's aesthetic.

Mulching, an age-old gardening practice, emerges as a reliable ally in the campaign for water conservation. By enveloping the soil with a protective layer of organic mulch, such as straw or wood chips, you effectively reduce evaporation rates and maintain ideal soil moisture levels. This practice ensures longer periods of hydration for your garden while reducing the need for frequent watering. The added benefit is the gradual improvement of soil health over time. Alternatively, inorganic options like gravel or pebbles might catch your fancy, offering a sleek, modern aesthetic and keeping both weeds and moisture levels in balance.

Amplifying your efforts through community involvement can have a dramatic impact on your conservation journey. Consider joining or even initiating local gardening groups that are driven by the shared pursuit of sustainability. Such initiatives foster community spirit and a collective sense of accountability. Within these groups, you may discover fellow gardening enthusiasts who share your passion for environmentally friendly practices and can exchange invaluable tips and resources with you. Community gardens are often epicenters of sustainable practices, featuring shared composting facilities and rainwater collection systems, contributing to collective conservation aims while encouraging social connectivity.

Adopting these water conservation strategies not only serves to enrich the environment but also enables the cultivation of a more sustainable garden. Through the embrace of these thoughtful practices, your garden can thrive without demanding excessive water usage. This harmonious balance reflects an alignment with both the garden's needs and the Earth's resources. Your garden then becomes a sanctuary symbolizing beauty and responsibility, embodying the nurturing of plants and the earth alike.

As you explore and incorporate these techniques, you'll soon come to realize that conservation transcends simply saving water. It morphs into the creation of a resilient ecosystem capable of flourishing with minimal human intervention. By embracing these methodologies within your gardening routine, know that each step you take is a contribution to the well-being of your plants and our planet.

TROUBLESHOOTING WATERING ISSUES: OVERWATERING AND UNDERWATERING

Navigating the balance between overwatering and underwatering can feel like learning a new language, but it's essential for the health of your container plants. Overwatering can lead to root rot, a condition where the roots can't function properly due to excess moisture. This often manifests as yellow, droopy leaves that paradoxically make the plant appear thirsty. On the flip side, underwatering causes the soil to dry out and pull away from the container's sides, while leaves become dry and brittle. Both scenarios stress the plant, but with careful observation and adjustment, you can learn to maintain the perfect moisture level for your container garden.

Addressing these watering missteps, whether rooted in excess or scarcity, requires decisive and thoughtful action. Should overwatering be the primary culprit, transforming your pot's soil blend is essential. Incorporate elements such as perlite or sand to enhance drainage, aerating the soil to prohibit water accumulation. Introduce these amendments gradually, adjusting the soil's composition until it fosters a balance between moisture retention and aeration. For those chronically underwatered plants, establishing a habitual watering routine is crucial. Employ the simple yet effective method of soil moisture testing—gently insert your finger into the soil to about an inch depth, evaluating its need for water. If the soil clings with moisture, resist the urge to water; if not, provide your plant the hydration it craves. This finger-test, casual yet immensely effective, allows you to tailor the water schedule specifically to each plant's needs, recognizing that one schedule does not fit all.

The adverse effects of persistent watering issues extend beyond the superficial aesthetics of drooping leaves. Unchecked, they can severely impact a plant's vitality, reducing its ability to bloom or bear fruit. Plants under stress from inadequate watering are akin to warriors weakened in the heat of battle, their defenses lowered, rendering them susceptible to pest invasions and diseases. This vulnerability can be likened to the human experience of battling illnesses while dehydrated—the body's barriers compromised, its fight significantly harder. Providing plants with an optimal balance of care—moisture, air, and nutrients—ensures they maintain their inherent protective abilities, warding off potential invaders.

To prevent future watering dilemmas, adopt the habit of vigilant monitoring. Regular assessments of soil moisture can preempt many common issues before they manifest. Develop the practice of adjusting your watering strategy as needed, an informed response to the subtle shifts in environmental conditions and plant growth. Maintaining a detailed watering log can be immensely beneficial—record each watering event, noting time, quantity, and any observed plant responses. Over time, this simple practice becomes a goldmine of data, revealing patterns and facilitating informed decisions to tweak and improve your care regimen.

As we draw this chapter to a close, we appreciate the intricate role that proper watering plays in successful container gardening. Beyond mere hydration, it involves creating conducive conditions where plants can flourish unabated. Mastering these watering techniques translates into healthier, more resilient plants, resulting in a thriving garden that resists pests, disease, and unfurls magnificent blossoms and fruits. Armed with these strategies, you are well-prepared to nurture a lush container garden, transforming your environment with beauty and vibrancy.

Next, we will delve into the art and science of designing your container garden layout for visual and functional brilliance. With your newfound understanding of watering behind you, embark on this next journey—curating a living masterpiece that truly reflects your personal style and creative vision.

CHAPTER 4
DESIGNING YOUR CONTAINER GARDEN

VERTICAL GARDENING: UTILIZING HEIGHT IN SMALL SPACES

Picture yourself standing on your tiny balcony, surrounded by lush greenery that climbs upwards, ambitiously defying gravity and transforming what was once an empty, unembellished wall into a lively vertical paradise. Vertical gardening represents not merely a practical solution for spatial constraints but also an artistic endeavor that adds rich layers of life and aesthetic beauty to your garden. Growing upwards allows you to maximize the potential of your available footprint, thereby making room for a greater abundance of plants than you might have envisaged possible. This method not only optimizes your usage of space but drastically enhances the visual appeal of the area by introducing depth and texture to your surroundings. The vertical element inherently draws one's eye upward, bestowing a sense of grandeur and expansiveness even in the smallest of spaces, creating an optical illusion of a more substantial garden.

Example of a vertical garden

Moreover, the benefits of vertical gardening extend beyond mere aesthetics. Elevating plants away from the ground reduces their exposure to soil-borne diseases and pests, allowing for better airflow around the foliage and stems of the plants. As a result, your plants can grow healthier and more robustly. Furthermore, the act of harvesting becomes a delightful breeze as fruits and vegetables conveniently dangle within easy reach, sparing you the need to crouch, bend, or stoop awkwardly. Popular systems for vertical gardening include trellises, wall planters, and hanging baskets.

Each presents distinct advantages suitable for varying space constraints and specific gardening needs.

The process of constructing vertical supports can be an immensely rewarding DIY project, or it might simply be an effortless trip to your local gardening store, contingent on your preference. For those inclined towards the handiwork of do-it-yourself projects, bamboo stakes, creatively bound with twine, make for excellent trellises. They stand firm, offer a natural aesthetic, and are wonderfully affordable. Additionally, recycling materials, such as old ladders or wooden pallets, can stimulate creativity and provide unique plant supports. Store-bought options, on the other hand, afford considerable convenience and a wide array of choices; wall planters, made in metal or plastic, simplify installation and maintenance very efficiently, serving well for herbs or flowers requiring shallow soil.

When choosing plants for a vertical garden, consider those species that naturally climb or trail. Ideal candidates include climbing plants like peas and beans, as they naturally seek upward support, making them organically suitable for trellises, and they produce substantial, fulfilling harvests. Their tendrils adeptly latch onto structures, which facilitates an easy training process along trellis work. Trailing plants such as nasturtiums spill forth vibrant blooms over the edges of planters, bringing splashes of vivid color and delightful charm, all while requiring minimal maintenance yet providing high-impact, visual results.

Maintaining a vertical garden calls for thoughtful attention to detail, but is far from overwhelming or daunting. Regular pruning is instrumental, as it promotes upward growth and aids in sustaining a neat and structured appearance. It is vital to remove any damaged or crowded leaves to improve light penetration, enhance air circulation, and foster healthier growth. Proper irrigation is crucial; ensure consistent water distribution through tools like drip systems or soaker hoses that deliver moisture directly to the roots, optimizing water usage and plant health. It's essential to frequently monitor soil moisture, as elevated containers typically dry out faster than ground-level ones, requiring occasional adjustments to watering practices.

INTERACTIVE ELEMENT: VERTICAL GARDEN PLANNING EXERCISE

Engage in an exercise where you take time to sketch out your envisioned vertical garden layout. Consider your available wall space, railings, and other spots amenable to vertical structures. Think through your plant choices based on sunlight exposure—select climbers for those bright, sunny areas, while opting for shade-tolerant trailers suited for shadier spots. This brainstorming and sketching activity will help you visualize your space's full potential, enabling you to make more informed and strategic plant selections that best suit your garden environment.

Vertical gardening is an exploration of endless possibilities when it comes to creativity and productivity. It represents a novel approach of reimagining your space and challenging the conventional paradigms of gardening. Embrace the vertical dimension and witness as your garden experiences unprecedented growth and beauty, flourishing like never before, creating an environment that is both functionally optimal and visually stunning.

CREATING EYE-CATCHING ARRANGEMENTS: COLOR, TEXTURE, AND FORM

Crafting a visually stunning container garden involves more than just planting your favorite flowers. It's about understanding basic design principles, which transform your space from ordinary to extraordinary. Color, texture, and form are your allies in this artistic pursuit. Let's start with color. Imagine the color wheel as your palette. Colors opposite each other, like blue and orange, create vibrant contrast and energy, while colors adjacent to each other, such as red and pink, offer harmony and subtlety. These combinations can evoke different moods. Warm colors like reds and yellows energize, while cool hues like blues and purples provide calm. By playing with these shades, you can shape your garden's atmosphere to reflect the feeling you want to evoke.

Texture adds another layer of interest to your arrangements. Mixing textures is like combining different fabrics in fashion; it creates depth and intrigue. Pair coarse-textured plants like lamb's ear or dusty miller with finer textures such as ferns or grasses. This contrast allows each plant to stand out, enhancing its individual characteristics. Texture isn't just about leaves; it includes stem shapes and the surfaces of flowers, too. Variegated leaves, which display multiple colors or patterns, are another way to introduce texture and visual interest. They break up monotony and draw the eye, adding complexity without overwhelming the arrangement.

Effective plant combinations bring these elements together seamlessly. Consider mixing bold foliage with delicate blooms. For instance, large-leaf hostas paired with dainty impatiens create a striking balance between strength and grace. Variegated leaves can be the showstopper in a container filled with simpler plants. Imagine a pot where the star is a variegated coleus surrounded by soft white petunias. The leaves' patterns catch the light differently throughout the day, offering dynamic visual shifts as time passes.

Arranging containers for maximum impact involves more than just choosing plants; it's about how you display them. Creating focal points is key. Use height variations to draw attention and guide the eye naturally through the space. A tall ornamental grass in a central container can act as an anchor, while smaller pots filled with mounding flowers fan out around it. Group containers with similar themes for a cohesive look, or mix styles for an eclectic feel. Consider using containers of varying sizes and shapes to add rhythm and flow to your display.

Seasonal changes offer opportunities to refresh your garden's appeal. Introducing seasonal annuals provides bursts of fresh color throughout the year. In spring, try planting pansies for early cheerfulness, followed by summer's bright zinnias. As autumn sets in, replace these with chrysanthemums for rich, warm hues that echo the season's natural palette. Rotating plants keeps your garden dynamic and ensures that there's always something new to catch the eye. It also allows you to experiment with different plant combinations each season.

Finally, remember that gardening is an ongoing conversation between you and your plants. You learn their language as you observe how they respond to different conditions and combinations. Adjust and adapt based on what you see—some plants might thrive together while others need more space or different care. Your garden becomes a reflection of your evolving tastes and understanding, a living canvas that changes with each choice you make.

Incorporate these principles into your container gardening practice to create arrangements that are not only beautiful but also meaningful. Each element—color, texture, form—contributes to a tapestry that speaks to both the heart and senses, inviting you to engage more deeply with your environment.

DIY CONTAINER PROJECTS: UPCYCLING AND CREATIVE SOLUTIONS

Envision looking around your living space, identifying unused items, and picturing them reborn as distinctive planters, each boasting its own unique appeal. Upcycling turns mundane objects into eye-catching elements, infusing both character and eco-friendliness into your gardening space. By repurposing different items, you not only give them a new lease on life but also reimagine their purpose and value creatively.

Let's first consider old tires. Often discarded without a second thought, these resilient rubber structures hold potential far beyond their original use. When stacked or painted, they can become bold, eye-catching planters that draw the admiration of onlookers. The inherent sturdiness of tires makes them robust containers suitable for accommodating larger plants or even small shrubs. By drilling several drainage holes and adding vibrant paint, these tires transform into colorful statement pieces, breathing new life into otherwise mundane corners of your yard or balcony. The durability of the tire material ensures these planters withstand the elements, offering longevity and enduring appeal for your green companions.

Wooden pallets offer another ingenious avenue for crafting vertical gardens. These commonly available structures can be repurposed into a lush tapestry of greenery mounted against your wall. To begin, sand them down to remove splinters and apply a weatherproof sealant, ensuring their outdoor resilience. Attaching small pots or fabric pockets along the slats, you create a tapestry of foliage, with each plant cascading downwards in a visual feast. This vertical arrangement not only optimizes space but also adds an inviting rustic charm to any environment. The natural texture and warmth of wood, combined with the vibrant life of herbs or flowers, create a harmonious and organic visual display that rejuvenates the senses.

For enthusiasts who enjoy a hands-on project, painting and sealing wooden crates can be an engaging and rewarding task. The key is to choose colors that harmonize with your garden's existing palette, employing non-toxic paints to guarantee safety for plants and the environment alike. Once painted, applying a waterproof sealant offers protection against moisture infiltration, extending the life of these delightful planters. The versatility of these crates adds layers of dynamism to any garden layout—stack them, line them up, place them at angles, or even hang them horizontally. This approach creates varying heights and structures, adding depth and interest to your garden's overall visual scheme.

Tin cans, often overlooked, hold potential for delightful transformation into quaint herb gardens with just a sprinkling of creativity. First, clean and strip the labels from the cans, then persistently use a hammer and nail to poke drainage holes in the bottom, ensuring proper water flow. Adding a coat of paint or decorative paper wraps elevates these simple objects into eye-catching containers, perfectly suited for windowsills or small ledges. Herbs like basil, mint, or rosemary thrive in these confined spaces, enjoying the warmth and nourishment provided within indoor environments. The beauty of a personalized container lies in its ability to reflect your unique style while engendering a positive environmental impact.

Finding materials for these projects requires little more than a keen eye and resourcefulness. Thrift stores offer a treasure trove of potential planters, each holding stories of past lives waiting to be discovered—old teapots, metal buckets, and woven baskets can all find newfound purposes within your garden. By joining local buy-nothing groups on social media platforms, you connect with a community of like-minded individuals eager to exchange items for free. Such exchanges open yet another avenue for sourcing materials at no cost, emphasizing community spirit and shared resourcefulness.

Engaging in upcycling projects fosters an essence of creativity, spurs sustainable living, and results in truly unique containers. Even the smallest space, with the aid of resourceful imaginations, transforms into a vibrant garden brimming with personality and character. By repurposing items in inventive ways, you not only aesthetically enhance your garden but also weave an intricate story —one of innovation, environmental care, and personalized flair.

DESIGNING WITH SUCCULENTS: LOW-MAINTENANCE BEAUTY

Step into a garden oasis that feels serene and tranquil, where every plant showcases resilience and adaptability. In this space, succulents shine as the stars of container gardening. They captivate both beginners and seasoned gardeners with their charm and robustness. Succulents are magical for their ability to thrive with minimal care, making them a perfect match for individuals with busy lives who might not always remember to water their plants. Their natural drought resistance means they're incredibly forgiving, ideal for those starting their gardening journey.

The dizzying variety of succulents further enhances their allure. From the soft, velvety texture of echeveria's intricately layered rosettes, which resemble delicate floral sculptures, to the robust, spiky allure of aloe with its medicinal prowess, succulents offer an eclectic mix of forms and hues. This diversity means you have an immense palette from which to craft a garden that isn't just visually enchanting but is a wholly personal expression of your aesthetic preferences and creativity.

When creating succulent arrangements, think of yourself as an artist with living plants as your medium. Each succulent is like a brushstroke, contributing to the overall beauty of your piece. For a striking composition, mix plants of different heights and shapes. For example, combine the tall, slender snake plants with the short, rounded hens and chicks. This variety creates an engaging visual journey. Adding decorative gravel or stones not only improves drainage but also enhances the aesthetic, giving your arrangement a refined, natural look. These touches mimic the succulents' native habitats, making your arrangement appear as though it's a slice of a sun-kissed desert brought into your home.

Caring for these resilient beauties is delightfully straightforward, yet adhering to a few key guidelines will ensure that your succulents maintain their vibrant health and lush appearance. Paramount among these is the selection of well-draining soil. Traditional potting soil is notoriously moisture-retentive and thus unsuitable; instead, choose a sandy, gritty mix, potentially augmented with additional elements like perlite or pumice, specifically designed for succulents and cacti. This eschews the peril of waterlogged roots, a common precursor to root rot—the archnemesis of succulents. Moreover, approaching watering with a philosophy of minimalism is essential: allow the soil to dry completely between drenchings, replicating the natural ebb and flow of dry conditions followed by infrequent, yet substantial, rainfall.

Succulents transcend the conventional confines of pots and planters, offering innovative display opportunities that infuse your living spaces with a dash of creative flair. Imagine the aesthetic possibilities of hanging succulent gardens, which are ideal for compact areas where horizontal surface area is scarce. Picture a vertical frame transformed into a living artwork, where vibrant succulents spill organically, evoking the beauty of wild cliffs or arboreal habitats. For a seasonal touch, a succulent wreath ingeniously offers a festive yet enduring alternative to traditional decor. By arranging tiny succulent cuttings within a circular design, supported by a sturdy wireframe and lush moss, you foster a dynamic centerpiece that morphs and grows over time, imbued with rustic charm and modern elegance.

The intrinsic charm of succulents is underscored by their effortless versatility and adaptability, rendering them the perfect choice for those desiring elegance without the encumbrance of intensive caretaking. The myriad forms, textures, and colors of succulents present you with limitless opportunities to unleash your creativity, enabling you to infuse your personal style into your environment while nurturing the serene tranquility that nature offers. Whether you're embarking on intricate crafts or venturing into novel displays, succulents provide an expansive canvas upon which your imagination can wander and innovate unfettered.

CREATING A POLLINATOR-FRIENDLY GARDEN IN CONTAINERS

Picture a vibrant garden, alive with the gentle hum of bees and the delicate flutter of butterfly wings; colorful blossoms sway gently in the breeze, painting your space with vivid hues and lively movement. This idyllic tableau isn't just a scene from a nature documentary—it's a pollinator-friendly garden right on your balcony or patio, crafted with intent and care. Pollinators, which include a diverse array of bees, butterflies, and even beetles, play an unquestionably crucial role in maintaining our ecosystem's integrity by facilitating the reproduction of plants through the transfer of pollen. Without these industrious creatures, our gardens, orchards, and even the broader food supply chain would face dire consequences. By enticing bees and butterflies to make your container garden their home, you not only enhance the beauty and productivity of your plant life but also make a valuable contribution to overall ecosystem health. These petite yet mighty creatures are pivotal in enhancing urban biodiversity, particularly in dense city areas where green spaces are often limited. They form essential pockets of life that provide sustenance not just to plants but to various wildlife forms intricately connected to these biological networks.

Choosing the right plants is paramount when it comes to captivating these industrious and beneficial insects. Take, for example, the enchanting lavender; its soothing, aromatic scent and delicate purple blooms act like a siren call for bees. Lavender flowers are rich in nectar, providing an irresistible treat that draws bees to your garden. Similarly, borage, with its striking star-shaped blue flowers, not only attracts bees with its nectar but also captivates the human eye with its vigorous and lively hue. For enthusiasts eager to support the mesmerizing monarch butterfly, the inclusion of milkweed in your garden is absolutely vital. Monarchs, in their enigmatic lifecycle, lay their eggs on milkweed plants, while their emerging caterpillars depend exclusively on the leaves for nourishment. This plant, therefore, not only sustains the monarch butterfly population but also infuses your space with a touch of wild, untamed beauty, reminiscent of vast meadows and thriving woodlands.

Designing a pollinator-friendly layout requires foresight and a holistic approach beyond merely selecting plants. It's about creating an ecosystem within your containers that invites, welcomes, and sustains tiny creatures. Start by layering your plants to provide a variety of access points that cater to different foraging habits. Taller plants, such as commanding sunflowers or majestic hollyhocks, offer shelter and strategic landing zones, while shorter plants like lavender and borage facilitate easy nectar access. This thoughtful plant layering echoes the natural stratification found in thriving landscapes, stimulating pollinators to explore various plant heights in their quest for sustenance. Integrate essential water sources as well—small, shallow dishes filled with pebbles can suffice, allowing pollinators to safely drink without the risk of drowning. These attentive touches transform your garden into a welcoming haven for visiting insects.

Maintaining a thriving pollinator habitat demands careful and thoughtful care. First and foremost, eschew the use of chemical pesticides that can inadvertently cause harm to pollinators. Numerous pesticides are toxic to bees and butterflies, disrupting their populations and hindering their pollination capabilities. Instead, embrace natural pest control methods that safeguard both your cherished plants and the vital insects that play their part in keeping your garden alive. Regular deadheading is a crucial ongoing task as well—by diligently removing spent blooms, you promote continuous flowering, ensuring that your garden remains an alluring spot for pollinators seeking nectar.

REFLECTION SECTION: POLLINATOR OBSERVATION JOURNAL

Embark on a reflective journey by starting a journal dedicated to observing the pollinators visiting your garden. Give notice to the multitude of species that grace your space, taking note of their preferred plants and any noticeable changes over time. This practice not only elevates your awareness of the indispensable role these creatures play but also deepens your understanding of how your garden nurtures and supports them. In time, this journal may become a treasured record of your garden's vibrancy and life.

Creating a pollinator-friendly garden transcends mere aesthetics; it is an invitation to give back to nature while immersing yourself in the breathtaking wonders of wildlife right at your doorstep. Watching bees buzz purposefully around lavender or delighting as butterflies perform an aerial ballet on milkweed evokes a profound sense of harmony and fulfillment that expands the joys of gardening into a form of stewardship and care for our shared environment. By consciously making these insightful choices, you play an integral role in sustaining the delicate balance of our natural world, all from the comfort and convenience of your own home.

INTEGRATING EDIBLES AND ORNAMENTALS FOR A FUNCTIONAL GARDEN

In envisioning a garden where aesthetics and functionality merge seamlessly, one imagines a place where beautiful blossoms and verdant leaves mix effortlessly with edible plants, creating a landscape that is not only pleasing to the eye but also fruitful in yield. This is the art and magic of mixed planting—a strategic approach to gardening that brings together the edible and the ornamental within your garden containers. This method does more than just optimize the limited space gardeners often face; it transforms your garden into a vivid and lively tapestry of colors, textures, and fragrances. By integrating both ornamental and edible plants, you cultivate an environment where every plant plays a crucial role in contributing to the overall utility and visual charm of your outdoor space.

Mixed gardens are significantly more than merely a visual delight. They offer numerous practical benefits that enhance and enrich your gardening journey. For instance, marigolds, when planted next to tomatoes, serve as natural pest deterrents, especially against aphids, thanks to their innate ability to repel various insects. This simple yet effective companion planting not only keeps your tomato plants healthier and free from pest infestations, but it also adorns your garden with brilliant bursts of orange, strikingly contrasting with the lush green foliage and deep red of the tomatoes. Likewise, the inclusion of basil next to petunias introduces a fresh and enticing aroma to your garden ambiance while simultaneously boosting its aesthetic appeal, courtesy of the vibrant hues of petunia blossoms. Furthermore, the robust fragrance of basil acts as a natural insect repellent, ensuring that your garden prospers with vitality.

Crafting and planning a mixed garden layout necessitates strategic thought and foresight to guarantee that each plant is situated in an environment conducive to its growth, receiving the optimal amount of sunlight and water. Begin by thoroughly assessing the individual needs of each plant species. More often than not, edible plants require full sun exposure to ensure ample production, whereas certain ornamental species flourish best in partially shaded conditions. Striking a balance between these requirements involves situating sun-loving varieties in the most sunlight-rich areas and allocating shaded spots for those that thrive with lesser direct sunlight exposure. Proper spacing is of paramount importance, as densely packed plants will vigorously vie for resources, potentially hampering their growth and development. Ensure there is adequate space for each plant to mature and expand without infringing upon the territory of nearby plants.

Sustaining a mixed garden effectively involves attending to the diverse and specific needs of different plant types. Edibles usually demand a higher intake of nutrients compared to ornamentals, making regular feedings an essential practice. Apply a balanced fertilizer consistently to maintain their health, vigor, and productivity throughout the growing season. Simultaneously, be vigilant with your ornamental plants, pruning them periodically to circumvent overcrowding and to improve air circulation. This essential maintenance not only elevates the visual appeal of your garden but also mitigates the likelihood of disease outbreaks.

REFLECTION EXERCISE: MIXED GARDEN PLANNING

Picture your ultimate mixed garden, taking into consideration which edibles and ornamentals you envision growing together. Draft a rough sketch of your layout, contemplating how these plants complement and enhance one another aesthetically and functionally. This introspective exercise in planning will aid you in strategically mapping out your garden, ensuring that each plant is positioned in a spot where it can thrive to its fullest potential.

Incorporating both edibles and ornamentals into your container garden offers a unique blend of aesthetic beauty and practicality. This harmonious combination means you can savor the joys of fresh produce ready for harvest while enjoying breathtaking blooms that augment your home's curb appeal. Through thoughtful planning and diligent care, one can cultivate a balanced and synergistic environment that offers delight and sustenance in equal measure, bestowing joy and nourishment upon your daily life.

As we conclude this chapter, let us contemplate how the blending of diverse plant types can enhance both the productivity and visual allure of your garden. It is about establishing harmony and synergy within your space, allowing you to embrace and relish the abundance of nature on multiple levels. In the subsequent chapter, we will explore methods and techniques to nurture healthy growth across a variety of plant species, ensuring your container garden remains vibrant and flourishing through the changing seasons.

MAKE A DIFFERENCE
WITH YOUR REVIEW

A SMALL ACT CAN GROW BIG THINGS

"The meaning of life is to plant trees under whose shade you do not expect to sit." – Nelson Henderson

People who help others without asking for anything in return are some of the happiest folks around. So here's a way you can make a difference!

There are tons of people out there who want to grow their own food or flowers but don't know where to begin. Sound familiar?

That's why I wrote *The Ultimate Beginner's Guide to Container*

Gardening. But here's the thing: I can only reach new readers with your help.

Most people look at reviews before picking a book. If you leave a review, even a short one, it could help someone else take that first step.

Your review could inspire...
• one more neighbor to grow their own fresh herbs,
• one more balcony to fill with blooming flowers,
• one more kid to fall in love with gardening,
• one more family to start eating what they grow,
• one more person to find peace through plants.

Writing a review takes less than a minute, but it could mean the world to someone else. **Just scan the QR or follow this link:**

https://www.amazon.com/review/review-your-purchases/?asin=B0FD478RY5

If you love helping people grow—plants and all—then you're my kind of person. Thank you so much for being part of this gardening journey.

Avery Sage

CHAPTER 5
DEALING WITH PESTS AND DISEASES

IDENTIFYING COMMON PESTS IN CONTAINER GARDENS

As you embark on your container gardening adventure, there's nothing more disheartening than discovering your flourishing plants under siege by tiny invaders. Picture this: you're admiring your vibrant greens, only to notice some leaves turning yellow and others looking a bit stunted. Your initial joy at seeing new growth is quickly replaced by a sense of foreboding—a sense that something is amiss. Welcome to the world of pest challenges in container gardening, where even the smallest critters can wreak havoc if left unchecked. Like silent marauders, these pests descend upon your garden, threatening the health and vitality of your cherished plants.

Consider two prevalent pests: aphids and spider mites. Aphids, small insects that feast on plant sap, vary in color, including shades of green, black, and white. These pests typically gather on new growth or beneath leaf surfaces, appearing as if marshaling for an onslaught. Aphids drain the sap from your plants, leaving a sticky substance known as honeydew in their wake. This residue is not only unsightly but also cultivates sooty mold, a fungal disease that cloaks leaves in a black, powdery film, obstructing vital sunlight. In contrast, spider mites are nearly invisible critters that create thin webbing around plants, causing leaves to exhibit a speckled or bronzed appearance when infested.

Recognizing the signs of these uninvited guests is crucial. Yellowing leaves, stunted growth, or visible insects on stems and leaves are clear indicators. Sometimes, the signs are subtle, whispering rather than screaming, yet they are no less significant. Inspect your plants closely—initial damage may not always be immediately obvious. That's where a magnifying glass becomes your best friend, helping you spot these tiny pests early. It illuminates a hidden world where these diminutive creatures carry out their destructive work. Turn over leaves regularly; pests often hide there, laying eggs or munching away unnoticed, like saboteurs in the night.

Monitoring your plants is like a regular check-up at the doctor's —essential for maintaining health. Think of it as a routine health examination for your verdant patients. Make it a habit to inspect your plants every few days, a gentle nudge that encourages you to stay connected with your garden. Use a magnifying glass for detailed inspections and gently turn leaves to check for eggs or larvae. This close inspection ensures that nothing escapes your notice. Early detection is key; catching pest issues early prevents them from spreading to neighboring plants and reduces the need for intensive treatments. Like a prudent guardian, you are preempting potential disasters, ensuring that your plants remain strong and resilient.

The importance of early detection cannot be overstated. It's the difference between a minor inconvenience and a full-blown infestation. By nipping problems in the bud, you protect not only the affected plant but also its neighbors from potential harm. This proactive approach saves you from resorting to heavy-duty measures later on. It's a strategic defense, a way of safeguarding your garden's future health and vitality.

Visual Element: Pest Identification Chart

Common Garden Pests

	Aphid	Tiny, pear-shaped insects (green, black, or yellow). Found clustered under leaves; they suck plant sap.
	Whitefly	Small, white, moth-like insects. Flutter up in clouds when disturbed; feed on underside of leaves
	Spider Mite	Extremely small, red or yellow. Leave behind fine webbing; cause stippling or yellowing of leaves
	Cabbage Worm	Green caterpillars that chew large holes in cabbage-family plants. Often blend in with
	Cutworm	Fat, gray or brown caterpillars. Hide in soil and cut young seedlings off at the base during

This handy reference can help you quickly identify any unwelcome visitors in your garden. Keep this chart in your gardening toolkit, your horticultural playbook against the forces of destruction. With a quick glance, you'll be able to assess threats and plan your response.

Remember, dealing with pests is part of gardening. It's a learning experience that strengthens your skills as a gardener. Like a chess player who learns from every move, you will become more adept at predicting and countering every threat. With vigilance and care, you can keep these pesky intruders at bay and ensure your container garden remains healthy and vibrant. Each challenge faced and overcome is a testament to your growth as a gardener, a reminder that the wonders of nature come with their own set of trials. But with perseverance and knowledge, you will prevail.

NATURAL PEST CONTROL: ECO-FRIENDLY SOLUTIONS THAT WORK

In the realm of container gardening, maintaining a healthy garden without relying on harsh chemicals is not only possible but preferable. Enter the world of natural pest control methods, where you can keep those pesky critters at bay while nurturing a safe environment for both you and the planet. Neem oil stands out as a helpful resource, particularly against soft-bodied insects like aphids. This oil, derived from the seeds of the neem tree, acts as a natural pesticide, disrupting the life cycle of these pests. When aphids feast on plants treated with neem oil, they ingest compounds that interfere with their ability to grow and reproduce. The beauty of neem oil lies in its specificity; it targets pests without harming beneficial insects like bees or ladybugs. It's a bit like having a security system that knows exactly who to let through and who to keep out.

Another effective method is using diatomaceous earth. This powdery substance is made from the fossilized remains of tiny aquatic organisms called diatoms. When sprinkled around your plants, it forms a barrier that deters crawling pests such as slugs and snails. As these pests crawl over the diatomaceous earth, the sharp edges of the particles damage their exoskeletons, leading to dehydration and ultimately their demise. It's a bit like walking barefoot over broken glass—unpleasant and something to be avoided. What makes diatomaceous earth particularly appealing is its non-toxic nature, ensuring that it poses no risk to your pets or family members. And since it's just crushed fossils, it's completely biodegradable, making it environmentally friendly as well.

Natural pest control methods offer numerous benefits beyond just a pest-free garden. They are safe for humans and pets, meaning you won't have to worry about harmful residues on your plants or in your home. Imagine being able to pluck a ripe tomato straight from the vine and pop it into your mouth without a second thought about chemical contamination. That's the peace of mind natural methods provide. In addition, these solutions contribute to environmental sustainability by reducing reliance on synthetic chemicals that can leach into the soil and waterways. You're not just protecting your plants; you're also playing a part in preserving the ecosystem.

Implementing these natural solutions is straightforward, even for beginner gardeners. To apply neem oil, mix two teaspoons of neem oil with a quart of water and a few drops of dish soap to help the mixture adhere to leaves. Pour this into a spray bottle and douse your plants, focusing on affected areas. The dish soap acts as an emulsifier, allowing the oil and water to mix effectively. For diatomaceous earth, simply dust it over leaves and around plant bases, forming an invisible fortress against intruders. This fine powder can be gently applied with a soft brush or shaken from a duster for even distribution.

While natural methods are highly effective, they do come with their own set of challenges. Ensuring thorough coverage is crucial for success; missing spots can allow pests to thrive unchecked. Take your time during application, turning over leaves and inspecting hidden crevices where pests might lurk. Reapplication is also necessary after rain or watering, as both can wash away protective layers. This might seem like extra work, but think of it as an opportunity to reconnect with your garden regularly. Each visit allows you to witness subtle changes and appreciate the resilience of your plants.

Remember, patience is key when using natural pest control methods. They may not offer instant results like chemical alternatives, but their long-term benefits far outweigh any initial delay. By opting for eco-friendly solutions, you're fostering a healthier garden environment and contributing to broader ecological health. As you continue cultivating your container garden, embrace these natural methods as an integral part of your gardening toolkit. They empower you to manage pests effectively while keeping your garden—and the planet—thriving.

As you explore these natural solutions further, consider documenting your experiences in a gardening journal. Note which methods work best for particular pests and how your plants respond over time. This reflection not only enhances your gardening knowledge but also creates a valuable resource for future reference.

PREVENTING AND TREATING FUNGAL DISEASES

Alright, let's delve into the intriguing yet somewhat vexing world of fungal diseases, the uninvited antagonists of the garden drama. These surreptitious adversaries have a knack for appearing when you're least prepared, crafting quite the mess as they spread their touch of mischief. Take powdery mildew, for instance—if you encounter a ghostly white, powdery substance enshrouding your leaves, it's the signature of powdery mildew. It's reminiscent of a careless dusting of flour across your precious foliage. Annoying, isn't it? Meanwhile, lurking beneath the surface is root rot, a stealthy menace that often slips beneath the radar until one day you find your plants drooping forlornly or, in worse scenarios, succumbing to demise. This typically ensues when roots languish in soil that's too saturated, leading to a slow, insidious decay.

To combat the threat of fungi in your garden, adopting smart strategies is essential. Start with considering the importance of air circulation, which is as vital for plants as it is for us. Overcrowded plants struggle to breathe, creating an ideal breeding ground for fungi. Ensure your plants have enough space by pruning them and spacing them properly to improve airflow. Next, assess your watering technique. Overhead watering may seem harmless, but it can leave leaves wet, inviting fungi to take hold. Instead, water the base of the plant, focusing on the roots where water is most needed. This keeps the leaves dry and helps deter fungal diseases.

However, apprehension can set in if fungi decide to pitch camp anyway. Stay calm; solutions abound. For powdery mildew—ever thought a simple kitchen staple like baking soda could be your savior? Mix one tablespoon of baking soda with a gallon of water, enhancing it with a few drops of liquid soap. Douse this mixture on the forlorn leaves witnessing the grim mildew, and prepare to observe its retreat. The magic lies in how baking soda modifies the pH balance on the leaf surfaces, crafting an inhospitable territory for further mildew invasion. On the other hand, when root rot rears its head, it's time for a garden surgery of sorts—a gentle uprooting and transplant operation. Delicately extract the plant from the soil confines, trim away the afflicted root portions, and repot it into fresh, well-draining soil. It's essentially a rejuvenation of your plant's foundational sustenance system.

Moreover, upholding a pristine garden ambiance is a formidable shield against the rampage of fungal diseases. Fallen leaves and debris aren't just untidy; they are potential hotbeds for spores, setting the stage for future infestations. Regularly purging these organic cast-offs can prevent fungi from securing an established foothold. Not to forget, sanitizing your gardening tools between uses is integral, particularly when dealing with contaminated plants. Tools can become unwitting couriers, ferrying spores and exacerbating the spread. A simple concoction of dish soap and water can transform your tools from accomplices back to allies.

Gardening is fundamentally a delicate balance, honed by minute attentions to detail. Fungal diseases may seem daunting upon initial acquaintance, yet through astute preventative measures and timely interventions, you can maintain a firm grip on their spread. Stay observant over the verdant lives you nurture, act swiftly when alterations are detected, and never hesitate to implement corrective measures. Every spot of powdery mildew or lamenting leaf presents itself as an invaluable learning curve, enabling the refinement of your enduring horticultural expertise.

REFLECTION SECTION: YOUR FUNGAL BATTLE PLAN

Take a moment for introspection. Secure a notebook or journal and pen down reflections regarding any fungal ailments you've confronted within your garden realm. Document the victims—specific plants impacted—and recount the strategies you adopted as a counter. Did the baking soda remedy find its application, or did you reassess your watering habits to accommodate the exigencies of plant health? Engaging in such reflective practice fosters a nuanced understanding and bolsters your growing confidence in grappling with fungal adversities.

Stay teeming with vigilance and anticipation, and recognize that tackling fungi is yet another integral chapter in the broader narrative of your gardening journey. Each defensive move and curative action reinforces your bond with the garden oasis you cultivate while solidifying your unwavering commitment as its guardian.

BENEFICIAL INSECTS: NATURE'S PEST CONTROL

In the vibrant and intricate world of container gardening, not all insects are adversaries. Imagine, if you will, a lively army of colorful ladybugs gracefully weaving through your plants, diligently devouring aphids and mites, much like a natural pest control brigade. These sprightly red warriors are more than just a visually delightful spectacle—they are a gardener's steadfast allies, capable of consuming hundreds upon hundreds of aphids throughout their busy lifetimes, boldly patrolling their domain. It's akin to having a specialized, devoted army perpetually on standby, poised and ready to defend your verdant sanctuary from those pesky intruders.

Pass to the valiant parasitic wasp, a tiny, often underrated hero of the garden. These minute warriors lay their inconspicuous eggs inside caterpillars and other garden pests, skillfully regulating their populations in a way that might, on the surface, seem somewhat macabre. However, this natural cycle cleverly sustains the delicate balance in your garden without the slightest reliance on chemical solutions. It's a marvelously efficient process that might seem gruesome but truly exemplifies nature's way of maintaining equilibrium and ensuring only the healthiest of ecosystems flourish.

CRAFTING AN ENTICING ENVIRONMENT

Attracting these benevolent insects transcends mere pest control; it embodies the creation of a thriving ecosystem in the very heart of your garden. These insects offer a sustainable method for managing pests, significantly reducing the need for synthetic chemicals that could damage both your plants and the broader environment. Consider a garden where diversity blossoms, each insect performing its individual role to uphold harmony within the ecosystem. Beneficial insects inherently support this balance by naturally curbing pest populations, thereby empowering your plants to grow more robustly and healthily. Moreover, their mere presence inspires other life forms to flourish, notably enriching your garden's biodiversity.

Envision your garden thriving with life, where each bug plays an integral role. To welcome these helpful creatures, certain adjustments can be effortlessly made. Planting nectar-rich flowers such as the humble alyssum acts as a natural draw for ladybugs and other beneficial insects. These blooms provide much-needed sustenance, enticing these insects to linger. Constructing insect hotels with simple structures filled with natural materials like twigs and bamboo can offer inviting shelters for beneficial bugs. Such cozy retreats allow bugs a tranquil sanctuary to lay eggs and weather any storms. By creating an abundance of food and shelter, you fashion an environment where beneficial insects readily flourish, offering continuous protection against pests.

CONCERNS AND BALANCES

Admittedly, you might worry about enticing too many bugs or upsetting the delicate balance within your garden. But rest assured, nature boasts an astounding talent for self-regulation. Inviting beneficial insects helps manage pest populations without letting any single species claim dominance. Should you notice an abundance of one insect type, observe your garden's dynamics closely. Often, multiple species naturally equilibrate populations over time. Be vigilant for signs of wholesome insect activity—look for ladybugs enjoying aphid meals or wasps busily flitting about their caterpillar hosts.

Addressing concerns about incorporating these creatures into your garden is surprisingly simple. Ensuring balance requires regular garden monitoring and only stepping in when absolutely necessary. If you discover that beneficial insects are thin on the ground, determine if they have ample resources—namely, food, water, and shelter—to thrive. Sometimes, modest alterations like increasing flowering plant varieties or minimizing pesticide usage can make a world of difference. Attentively observing your garden for prolonged periods fosters an understanding of its unique ecosystem and enlightens you on the optimum ways to support it.

EMBRACING THE UNSEEN GUARDIANS

Cultivating an environment that welcomes beneficial insects extends beyond mere plant protection; you are nurturing a miniature ecosystem that thrives on balance and diversity. These minuscule helpers toil tirelessly behind the scenes, ensuring your container garden retains its vitality and resilience. Embrace their formidable yet understated presence as a critical facet of your gardening odyssey—a fundamental element in crafting a vibrant and sustainable outdoor space.

The endeavor of gardening embraces the unexpected, finding splendor in nature's intricate tapestry of life. As you cultivate your container garden, cherish the knowledge that every ladybug and parasitic wasp is a steadfast ally in your mission to maintain a flourishing, pest-free haven. Their presence speaks to nature's remarkable capacity for balance and rejuvenation—a poignant reminder that oftentimes, the smallest creatures yield the most profound impacts on our gardens' health and well-being.

DIY REMEDIES FOR PESTS: RECIPES AND TECHNIQUES

Imagine stepping into your garden, feeling the warmth of the sun on your face, only to be confronted by an army of pests nibbling away at your hard work. Before you reach for commercial chemical sprays, consider some homemade remedies that are both effective and environmentally friendly. Garlic and chili spray is a powerful deterrent against aphids, those pesky little sap-suckers. To make this, blend two cloves of garlic with a few spicy chilis in a quart of water. Let it sit overnight, then strain into a spray bottle. The potent aroma of garlic and the fiery kick from chilis create an unappealing environment for aphids, sending them scurrying away.

Another tried-and-true method is a classic soap and water mixture. Perfect for dealing with soft-bodied insects, this remedy involves mixing one tablespoon of liquid dish soap with a quart of water. The soap disrupts the cell membranes of insects like spider mites, causing them to dehydrate and ultimately perish. Applying this solution with a spray bottle allows you to target specific problem areas without harming beneficial insects. These simple concoctions are easy to whip up with ingredients you likely have at home, making them a convenient first line of defense against garden invaders.

While DIY remedies can be a gardener's best friend, they aren't without their limitations. They work wonders for small-scale infestations but might struggle against more substantial pest populations. Frequent reapplication is often necessary, especially after rain or watering, as these natural solutions can easily wash away. Think of them as part of your regular garden maintenance routine rather than a one-time fix. Although they may require a bit more effort than commercial alternatives, the peace of mind knowing you're not introducing harsh chemicals into your garden is well worth it.

When applying these remedies, safety and plant health should be top priorities. Conduct a spot test on a few leaves before proceeding with full application, as some plants might be more sensitive to certain mixtures. This small step can prevent accidental leaf burn or other damage. Apply these solutions during cooler parts of the day—early morning or late afternoon—when evaporation rates are lower and plant stress is minimized. This careful timing ensures that your plants receive the full benefit of the treatment without unwanted side effects.

Experimentation and adaptation are key to mastering DIY pest control. Feel free to adjust concentrations based on your plants' sensitivity or combine remedies with other natural methods to boost effectiveness. For instance, adding a bit of vegetable oil to your garlic and chili spray can help the mixture adhere better to leaves, increasing its longevity. Keep in mind that each garden is unique, so what works for one might not work for another. Embrace this as an opportunity to learn more about your plants and tailor your approach to their specific needs.

If you're feeling adventurous, consider keeping a garden journal to track your experiments with different remedies. Document what works, what doesn't, and any tweaks you make along the way. This not only helps refine your pest control strategies but also provides valuable insights into your plants' responses over time. Your journal becomes an evolving resource—a personal guide that grows with each season and experience.

Incorporating DIY pest control methods into your gardening routine fosters a sense of empowerment and creativity. You're not just combating pests; you're engaging with your garden on a deeper level, learning its secrets and rhythms. Each recipe you try adds another layer to your understanding of the delicate balance between plants and their environment. With every successful application, you gain confidence in your ability to protect your garden naturally.

So go ahead—experiment, adapt, and discover what works best for your container garden. Embrace these DIY remedies as part of your gardening toolkit, knowing that you're contributing to a healthier garden and a healthier planet. With patience and persistence, you'll find yourself not only nurturing plants but also cultivating your skills and knowledge as a gardener.

CREATING A HEALTHY ECOSYSTEM: PROMOTING PLANT RESILIENCE

In the captivating interplay of nature, a balanced ecosystem assumes the starring role, much like the conductor of a symphony orchestrating harmonious notes. Picture your garden as a self-contained microcosm, a vibrant mini-universe where every leaf, insect, and water droplet plays a crucial part in nurturing a sustainable environment. This delicate web of interconnectedness is essential for naturally minimizing pest and disease pressures, allowing gardeners to reduce reliance on harsh, chemical interventions. By welcoming biodiversity into your garden, you encourage a multitude of life forms that coexist in balance, each contributing uniquely to the overall harmony. Envision incorporating a diverse mix of plants—from exquisite flowering species to lush leafy greens—crafting a kaleidoscopic tapestry that baffles pest populations and fortifies the ecosystem's resilience.

Maintaining soil health stands as another pivotal pillar of a resilient garden. A well-nurtured soil brims with beneficial organisms that enable plants to efficiently fend off diseases while absorbing nutrients. Adopting organic practices, such as composting and mulching, enriches the earth, transforming it into a bountiful banquet of nutrients for your plants. When your soil flourishes, your plants grow more robust and resilient, well-prepared to withstand the caprices of nature. Imagine the transformation as your plants flourish with vigor, drawing strength from the rich, living soil beneath them, intertwining roots with an intricate network of microscopic allies.

To bolster resilience against pests and diseases, initiate by meticulously selecting plant varieties renowned for their resistance. Many seed catalogs proudly showcase these traits, simplifying the process of making informed selections. Additionally, providing balanced nutrition through strategic fertilization holds equal importance. Like humans, plants thrive on a comprehensive and balanced diet. Utilizing slow-release fertilizers ensures they receive a steady influx of essential nutrients over time, reinforcing their natural defenses and improving their robustness. Imagine how, with each passing day, your plants become fortified warriors, equipped to repel potential threats.

Companion planting emerges as another hidden gem in your gardening toolkit. Through strategic pairing of plants, you can significantly enhance overall garden health while naturally deterring pests. Aromatic herbs such as basil and mint can cloak the scent of more vulnerable plants, effectively keeping pests at bay. Alternatively, employing trap crops such as nasturtiums can be a strategic addition to your container garden. These plants serve as sacrificial lures, attracting pests away from your valuable crops. By diverting the attention of insects, nasturtiums effectively safeguard your cherished plants, acting as a natural barrier against unwanted guests. This method not only reduces the need for chemical pesticides but also adds an aesthetic appeal to your gardening space with nasturtiums' vibrant flowers. This shrewd utilization of plant partnerships not only amplifies your garden's productivity but also markedly reduces dependency on chemical interventions. Visualize the bustling life in your garden, where symbiotic plant alliances quietly wage battles against invaders, fostering an organic, thriving ecosystem.

Engaging in regular observation and adaptation plays a critical role in sustaining a balanced ecosystem. Maintaining a garden journal allows you to chronicle changes, detail successful strategies, and document interventions. This valuable practice aids in maintaining a strong connection with your garden's needs, allowing you to decipher patterns over time. By attuning yourself to the seasonal shifts and plant responses, you can fine-tune your practices, ensuring your garden remains a verdant and lively tapestry throughout the year. Imagine yourself as a seasoned detective, piecing together clues in your garden journal, always ready to adapt and refine your strategies.

In conclusion, crafting a robust ecosystem within your container garden hinges on embracing biodiversity, nurturing soil health, selecting resistant plant varieties, and employing strategic companion planting. By cultivating such an environment, you foster a realm where plants flourish naturally with minimal intervention. Continuous observation and adaptation ensure you remain in sync with your garden's dynamic needs, promoting resilience against pests and diseases.

As we draw this chapter on confronting pests and diseases to a close, remember that every challenge presents an opportunity for growth and learning as a gardener. Your garden, like an artist's canvas, is an evolving masterpiece, constantly presenting unexpected discoveries.

CHAPTER 6
ADAPTING TO CLIMATE CHALLENGES

UNDERSTANDING YOUR CLIMATE: MICROCLIMATES AND PLANT HARDINESS

magine you're planning a picnic. You check the weather, pack accordingly, and hope for the best. Gardening is somewhat similar, but instead of a weather app, you have the fascinating concept of microclimates right in your backyard. Think of your garden as a patchwork quilt, where each square has its own unique climate. Maybe one corner is sunnier because it's next to a south-facing wall that soaks up heat, while another spot stays cooler in the shade of a tree. Recognizing these variations is key to selecting the right plants for each area.

Microclimates are tiny climate zones within your garden, and they can make a big difference in how your plants grow. A south-facing wall can create a warmer microclimate, perfect for heat-loving plants like peppers or tomatoes. These walls absorb sunlight during the day and release warmth at night, extending the growing season slightly and offering protection during cooler nights. In contrast, shaded areas under trees or beside buildings maintain cooler temperatures, which is ideal for plants that prefer less intense sunlight, like ferns or hostas. Understanding these nuances helps you tailor plant choices to each unique spot.

To truly master your garden's potential, knowing your plant hardiness zone is essential. The USDA Plant Hardiness Zone Map is a fantastic tool for this (https://phzm-prod.ars.usda.gov). It divides regions based on average minimum winter temperatures, helping you determine which plants can survive the winter in your area. But here's the twist: microclimates can exist within these zones, adding another layer of complexity. For example, if your home is in Zone 7 but you have a particularly sheltered spot, it might behave more like Zone 8. Recognizing these differences allows you to push boundaries and experiment with plants that might not typically thrive in your broader zone.

Once you've identified your local hardiness zone and microclimates, adapting your plant choices becomes an exciting game of strategy. In cooler, shaded spots, opt for shade-tolerant plants like lettuce or spinach that appreciate the respite from direct sunlight. These plants thrive without too much heat and can offer lush greenery even in the most tucked-away corners. On the other hand, sunny exposures call for heat-tolerant varieties such as rosemary or lavender, which bask happily in bright light and warm temperatures. By aligning plant preferences with their ideal microclimates, you can create a diverse garden that flourishes across all its varied sections.

To better understand and manage these microclimates, investing in a few simple tools can be invaluable. A basic thermometer helps track temperature variations throughout the day, revealing which areas stay consistently warmer or cooler. Humidity sensors provide insight into moisture levels, helping you adjust watering practices to suit each microenvironment's needs. These tools help you make informed decisions, ensuring each plant receives optimal care tailored to its specific conditions.

INTERACTIVE ELEMENT: CLIMATE MAPPING EXERCISE

Grab a notebook or start a digital journal and map out your garden's microclimates. Spend a week observing how sunlight moves across your space at different times of day. Note temperature fluctuations using a thermometer and identify any particularly dry or damp areas with a humidity sensor. Sketch these observations on paper or digitally to create a microclimate map of your garden. This exercise not only enhances your understanding but also informs future planting decisions by highlighting the diverse conditions within your garden.

Adapting to climate challenges involves more than just choosing the right plants—it's about embracing the dynamic nature of your garden's environment. By recognizing microclimates and understanding plant hardiness zones, you gain the confidence to make informed choices and create a thriving container garden that reflects both nature's diversity and your personal style.

PROTECTING YOUR PLANTS FROM EXTREME WEATHER: HEAT AND FROST

Picture a sweltering summer afternoon, and the gardens are wearing a shimmering, mirage-like haze under the relentless sun. The leaves droop tiredly, silently pleading for respite. It's time to intervene and grant your beloved flora some relief. An innovative and effective way to shield these plants from the scorching sun is by employing shade cloths. These lightweight fabrics act as saviors, gently filtering the sunlight that reaches your garden, subtly reducing the intensity and significantly keeping your precious plants cooler. Imagine draping them artfully over the garden during the peak heat hours of the day, establishing a protective sanctuary bathed in dappled light—a refuge from the merciless rays. This clever yet straightforward method helps prevent unsightly sunburn on leaves that could lead to more severe dehydration, thus lowering the risk of heat stress that might otherwise pummel your garden into wilted submission.

Furthermore, considering the tactile intimacy of mulching—a practice as old as gardening itself—imagine spreading a generous layer of organic mulch, such as straw, shredded bark, or compost around your plantings. This not only serves as a gentle covering but also as a natural temperature moderator. The mulch acts like an insulating blanket, preserving essential moisture within the earth's crust, making sure the soil temperature remains delightfully consistent despite varying atmospheric conditions. With these multifaceted strategies proactively in place, you're essentially fortifying your garden against the arduous summer heat without breaking a sweat, allowing you to enjoy the scene of life sprouting with vigor.

Fast forward to a chilly autumn night when you can envision frost stealthily creeping like a whisper, nipping at the delicate edges of leaves. Frost is indeed treacherous, often appearing out of nowhere and causing significant damage if you're caught unprepared. To combat its stealthy attacks, covering plants with frost blankets or cloches becomes an astoundingly effective defense mechanism. These covers are designed to trap the warmth emitted from the soil, thus encapsulating a cozy microclimate around your plants. Securing the edges is crucial to prevent any sneaky cold air from crawling underneath. For those plants kept in smaller containers, consider relocating them indoors or placing them in sheltered areas like a garage or an enclosed porch. Such a nimble relocation shields the more delicate plants from frost's icy clutches, therefore lending them a fighting chance to breathe life anew when the morning sun returns.

Indeed, sudden weather changes can occasionally catch even the most seasoned gardener off guard, with unpredictable temperature shifts posing a frequent threat. When confronted with such unexpected situations, having a calculated contingency plan is key.

Gardener foresight is invaluable, especially when garden screens or temporary barriers can act as vigilant windbreaks, protecting the plants from chilling gusts that mercilessly strip away accumulated warmth. Should temperatures plummet mysteriously overnight, a simple and effective trick is to water your plants in the evening. This subterfuge works because water naturally retains warmth, slightly elevating soil temperatures through the cold night and providing a semblance of warmth to sustain your plants.

Choosing climate-resilient plants is yet another brilliant way to strengthen your gardening tapestry against weather extremes. In blistering hot climates, consider incorporating heat-tolerant succulents like agave or aloe vera. These resilient beauties thrive with minimal water, happily basking in intense sunlight, rendering them perfect incumbents for sun-drenched locales. Conversely, if you reside in regions tainted by frost, selecting frost-hardy perennials, such as hellebores or Siberian iris, becomes imperative. These valorous plants can withstand cold incursions without wilting, bringing color and vibrancy even when the thermometer's mercury sinks.

Embracing climate-resilient gardening extends beyond mere plant selection; it embodies the nurturing of a resourceful, adaptable mindset to expertly navigate nature's unpredictable whims. As your green thumb gains experience and confidence, you'll soon learn to anticipate local weather patterns, adeptly responding to ever-changing conditions. In time, your garden will not only chirp as a testament to your toil and care but also reflect the resilient spirit of sheer adaptability amid nature's myriad challenges.

By incorporating these diverse strategies and choosing robust plant varieties, you will arm yourself with an arsenal of defenses to protect your garden from both the fierce summer heat and the biting winter frost. Your verdant oasis will thrive through seasonal shifts, offering perennial beauty and bountiful produce, notwithstanding nature's unpredictable temperament.

ADAPTING TO SEASONAL CHANGES: TIPS FOR YEAR-ROUND GARDENING

Think of your garden as a dynamic stage where plants become the protagonists of a fluctuating, ever-evolving performance, responding harmoniously to the seasonal symphony orchestrated by nature. By engaging in the practice of crop rotation, you invigorate this stage, ensuring it remains lively, fertile, and highly productive throughout the year. In the budding days of early spring, cool-season crops such as crisp lettuce, tender spinach, and sweet peas make their grand entrance. These resilient plants enthusiastically embrace the gentle chill of early growth, flourishing vibrantly before the summer sun intensifies its warmth. As the mercury rises, usher in the vibrant cast of warm-season vegetables, including juicy tomatoes, fiery peppers, and crunchy cucumbers. These sun-worshipping stars bask in the swelling heat, offering generous harvests amidst the sultry, sun-drenched months. The art of rotating plants not only enhances your garden's productivity but also plays a crucial role in mitigating soil nutrient depletion, effectively disrupting pest cycles, and encouraging a balanced ecosystem teeming with vibrant life.

Maintaining a garden that enchants year-round demands a touch of foresight paired with a pinch of artistic savvy. By thoughtfully incorporating evergreen foliage, you ensure a perpetual splash of color even during the bleakest days of winter. Stalwarts like boxwood or juniper stand guard, maintaining the vibrancy of your garden landscape as other plants retire from their seasonal duties. Concurrently, planting seasonal annuals such as cheerful pansies in the lively springtime or bright zinnias during the sunlit summer months ensures a steady influx of fresh blooms, continually refreshing your garden's dynamic palette. Selecting plants with varied blooming timelines injects a continuous rhythm of vibrancy and variety, seamlessly transitioning from one season's charm to the next. This curated approach ensures each season brings its own unique flair, transforming your garden into a perpetually appealing visual feast, adorned with colors and resplendent with life.

Extending the growing season stands as a cherished dream for gardeners, a pursuit achievable through the application of clever techniques and inventive solutions. Cold frames function as ingenious miniature greenhouses, artfully capturing sunlight and retaining heat, providing protection to early spring crops from the chilling grasp of frosty nights. Envision these structures as snug sanctuaries where your plants can continue to thrive, defying the brisk air that characterizes early spring. In a similar vein, row covers operate like comforting blankets, gently draped over your garden beds, extending the bounty of the fall harvest by safeguarding plants from cold wind gusts and light frost. Employing these valuable tools allows you to transcend the boundaries traditionally defined by growing seasons, granting you the pleasure of extending your enjoyment of homegrown produce for many months more.

INSIGHTFUL APPROACHES TO SEASONAL PLANNING

Seasonal planning emerges as an indispensable component of any successful gardening endeavor, necessitating the crafting of dedicated task lists tailored to each climate phase, thus elevating the organization and efficiency of your gardening chores. In the rebirth of spring, devotion is set upon preparing the rejuvenating soil and planting the hearty cool-weather crops. Summer ascends with tasks such as vigilantly weeding and adjusting the intricate irrigation systems to accommodate scorching days, thus maintaining steady hydrating love to counter intense evaporation. As fall approaches, it ushers in a period of reflection and preparation. It's a time to tidy up diligently, engaging in thoughtful end-of-season pruning and applying protective mulch to shield dormant plants through the icy grip of winter. Adjustments in watering and nourishing schedules are essential, synchronizing with the shifting seasons to ensure that each plant receives precisely the care it demands at any given time. For instance, during the towering temperatures and intense heat of summer, plants may require more frequent watering regimens or additional nutrients to bolster their rates of growth.

Gardening is a lively and interactive dance with the inexorable cycles of nature, where every season unveils novel opportunities for introspection, learning, and adaptation. Embracing these transformations with open-hearted innovation and understanding sustains your garden's vigor and vibrancy throughout the year. By adeptly rotating crops, selecting plant varieties that maintain perpetual interest, extending growing seasons thoughtfully, and planning activities around seasonal rhythms, you cultivate a resilient garden that thrives, robustly weathering every climatic shift.

The essence of gardening success resides in an attuned awareness of and synchrony with the natural world's rhythms, remaining poised and adaptable amidst nature's dynamic variations. It involves far more than merely sowing seeds; it's about nurturing an interlocking ecosystem that supports and sustains abundant life throughout the entire cyclical journey of the year. By meticulously planning ahead, coupled with doses of creativity and intuitive insight, your container garden blossoms into a perennial source of joy, beauty, and abundant harvest, transcending every seasonal boundary.

CONTAINER GARDENING IN ARID CLIMATES: XERISCAPING AND MORE

Step into a garden bathed in sunlight, a place where every plant tells a story of resilience and smart gardening. This vision captures the spirit of xeriscaping, a gardening philosophy that excels in dry, challenging climates by prioritizing water conservation. In areas where water is a scarce resource, xeriscaping offers a practical guide to crafting stunning, low-maintenance gardens that use water sparingly. The approach centers around choosing drought-resistant plants, the true heroes that withstand intense heat with minimal water. Imagine integrating succulents like jade plants or cacti into your container garden. Their thick, water-storing leaves make them ideal for thriving in hot conditions with little upkeep, embodying the essence of water-efficient gardening.

However, xeriscaping encompasses much more than just the strategic selection of plants; it embodies an artful approach to planting that reduces water use while magnificently maximizing aesthetic appeal. In a container garden, this translates into grouping plants with similar water needs together, ensuring efficient watering practices and minimizing waste. Imagine a thoughtfully arranged cluster of ornamental grasses and native perennials like lavender or yarrow. These plants not only thrive with minimal watering but also add rich texture and vibrant color to your garden palette. By thoroughly understanding the specific water requirements of each plant and positioning them accordingly, you craft a harmonious oasis that requires less from you but invites more from nature's bounty.

Choosing the appropriate soil and containers in arid climates can significantly amplify the success and sustainability of your xeriscaped garden. Terracotta pots, with their naturally porous nature, allow for natural moisture retention while crucially providing the essential airflow needed by plant roots. These timelessly crafted containers not only lend a rustic charm but also effectively support the overarching goal of reducing water loss through evaporation. Pair these pots with water-retentive soil amendments such as vermiculite or coconut coir, which are instrumental in maintaining optimal moisture levels. These materials form the backbone of an environment where plants can truly thrive even in dry conditions. By wisely investing in the right soil and containers, you lay the groundwork for a flourishing garden that admirably defies the often overwhelming challenges posed by an arid climate.

Watering in arid climates requires a degree of finesse and careful thoughtfulness. A drip irrigation system can become your best ally, delivering water directly to the plant roots where it's needed most, thereby minimizing waste and ensuring each precious drop is efficiently utilized. Furthermore, adopting a practice of deep but infrequent watering encourages roots to grow deeper into the soil, seeking out moisture reserves hidden below the surface. Think of it as teaching your plants to become self-sufficient explorers, venturing far and wide in search of sustenance. This approach not only conserves water but also builds resilience in your plants, preparing them to withstand the harshest conditions.

When considering plant selection, cacti and succulents naturally top the list for their minimal water needs and evocatively striking appearances. Imagine a stunning collection of echeverias with their perfectly sculpted rosette forms or a stately row of agaves standing tall and proud. These plants embody the true spirit of xeriscaping, thriving with minimal care while showcasing nature's unparalleled artistry. Beyond these classic options, consider incorporating native plants that are deftly adapted to your local conditions. Over time, they've evolved to withstand the specific challenges of your climate, making them reliable and invaluable additions to any arid garden. Picture wildflowers like California poppies or desert marigolds adding extraordinary splashes of color and texture to your landscape.

Creating a successful container garden in an arid climate is both an artful endeavor and a scientific pursuit. It involves a profound understanding of the unique challenges posed by water scarcity and employing innovative strategies that align harmoniously with nature's rhythms. By fully embracing xeriscaping principles, you invite sustainability into your garden, crafting a space that flourishes magnificently against the odds. From meticulously selecting drought-tolerant plants to expertly choosing the right soil and containers, every decision contributes to a thriving oasis that celebrates resilience, beauty, and creativity. With thoughtful planning and a touch of creativity, your container garden can become a vibrant testament to what's possible even in the harshest environments, dazzling the senses and nurturing the soul.

COPING WITH HUMIDITY: PLANT CHOICES AND CARE STRATEGIES

Gardening in humid climates presents its own unique set of challenges, akin to navigating a thick, damp fog. High humidity can create a breeding ground for fungal diseases, turning your once-thriving garden into a haven for mold and mildew. Imagine the frustration of discovering powdery mildew creeping over your beloved plants, or the disappointment of seeing leaves yellow and wilt under the weight of excess moisture. Managing water levels becomes a delicate balancing act, as too much moisture leads to root rot, while too little can leave your plants parched. The humidity blankets everything, making it tough to maintain the ideal conditions for your green companions.

To combat these challenges, selecting humidity-tolerant plants can make a world of difference. Tropical plants like ferns and orchids thrive in humid environments, their lush foliage and exotic blooms adding a touch of paradise to your garden. These plants have adapted to high humidity, drawing moisture from the air to stay vibrant and healthy. Similarly, vegetables such as okra and peppers perform admirably in these conditions. They relish the warm, moist air and produce bountiful yields when given the care and attention they need. By choosing plants that naturally flourish in humidity, you set your garden up for success.

Managing moisture levels in a humid climate requires a few clever strategies. Increasing air circulation with fans is a simple yet effective way to keep air moving around your plants, reducing the risk of fungal growth. A gentle breeze can do wonders for keeping leaves dry and preventing mold from taking hold. Additionally, using well-draining potting mixes is crucial to prevent water from lingering too long around your plant's roots. These mixes typically contain materials like perlite or sand to facilitate drainage, ensuring that excess water escapes quickly and efficiently.

Humidity often invites unwanted guests—pests that thrive in damp environments. Regular inspection of your plants becomes essential, as early detection can prevent a minor issue from snowballing into a full-blown infestation. Keep an eye out for signs of mold or mildew on leaves and stems, as these can indicate both fungal problems and potential pest activity. Implement preventative pest control measures by maintaining cleanliness in your garden space. Clear away fallen leaves and debris that may harbor pests or diseases, and consider using natural remedies like neem oil or insecticidal soap to deter invaders.

It's also worth noting that some pests, like aphids, are particularly fond of humid conditions. These tiny insects suck sap from plants, weakening them over time. To combat this, ensure your plants are well-nourished and healthy, as robust plants are less likely to fall victim to pest pressures. Consider introducing beneficial insects like ladybugs or lacewings to your garden; they naturally prey on aphids and other common pests, keeping populations in check without the need for harsh chemicals.

For gardeners dealing with relentless humidity, finding harmony between moisture and plant health is key. Experiment with plant placement, ensuring good airflow between containers and within the garden space itself. This simple adjustment can reduce humidity-related issues significantly. Embrace trial and error as you navigate the complexities of humid gardening—each lesson learned brings you closer to mastering these unique conditions.

To further aid in managing high humidity, consider incorporating raised containers or vertical gardening techniques. Elevating plants can improve air circulation around them, reducing the chance of fungal infections. These strategies not only tackle humidity but also make efficient use of space, adding vertical interest to your garden layout.

Incorporating reflective surfaces like light-colored gravel or stones around potted plants can also help mitigate humidity effects by reflecting sunlight and heat, thus moderating soil temperatures. This subtle adjustment contributes to a more balanced environment where plants can continue thriving despite the challenges posed by excessive moisture.

Gardening in humid climates may require some extra effort, but with thoughtful plant selection and strategic care practices, it's entirely possible to cultivate a flourishing container garden. Embrace the lushness that humidity can bring while staying vigilant against its challenges, ensuring your garden remains a vibrant oasis amidst the dampness.

WINTERIZING YOUR CONTAINER GARDEN: PREPARING FOR COLD MONTHS

As the curtain of winter unfolds with its chilly embrace, the task of preparing your container garden for the upcoming cold months takes center stage. Picture the process as tucking your beloved plants into a snug, cozy blanket, ensuring their comfort during the long, dormant stretch until the warmth of spring reappears. One particularly effective method of doing this is by insulating pots with materials such as bubble wrap or burlap. This seemingly simple yet profoundly impactful act provides a much-needed buffer against the relentless bite of cold, especially for containers crafted from materials like ceramic or clay, which are prone to cracking under freezing conditions. By carefully wrapping the chosen insulating material around the pot and securing it with twine, you create a snug cocoon that serves to trap warmth, lovingly shielding your plants from the harshness of winter.

Additionally, in nature, we observe many species coming together for warmth, and your plant containers can do the same.

Grouping them will, much like the instinctive huddle of penguins to preserve heat, allow your plants to share warmth and decrease their exposure to frost's icy grip. As temperatures continue to plunge, consider offering a refuge by moving some of your cherished plants indoors. This transition, however, requires a deft touch to ensure a seamless adjustment to their new indoor environment. Begin by gradually acclimating them to indoor conditions. You can bring them inside for short bursts, incrementally expanding that time over a one or two-week period. This gradual exposure assists them in adapting to the variances in light levels and humidity indoors. Ensuring appropriate light exposure is of the utmost importance, so position the plants near bright windows where they can bask in natural sunlight. If this is not feasible, strategically deploy grow lights to complement natural sunlight. Indoor air, particularly when the heating systems hum into action, can be drier, leading to potential stress for the plants. Thus, consider utilizing a humidifier or placing water trays around plants to uphold adequate moisture levels.

In winter's embrace, plants tend to enter a dormancy phase, demanding care routines that diverge from those during the active growth periods. A pivotal task during this stage is the careful reduction of watering frequency. With growth slowing to a gentle halt, the plants inherently require less water, and overwatering can be a slippery slope to root rot. Regularly feeling the soil's moisture status provides a useful guide, watering only when the soil feels dry about an inch beneath the surface. Another essential task is the meticulous pruning of any dead or damaged foliage. This invigoration helps channel the plant's energy towards bolstering its healthy components while simultaneously warding against potential pest infestations.

Winter is not synonymous with neglect; rather, it invites regular checks to ensure your plants maintain their vigor. Keep a vigilant watch on indoor plants, scanning for any signs of pest activity, such as the unwelcome presence of spider mites or aphids. These minute disrupters can stealthily infiltrate indoor sanctuaries and unleash havoc if not promptly addressed. By regularly inspecting leaves and stems, and gently wiping them with a damp cloth if necessary, you can maintain their health and vitality. As fluctuating temperatures weave through the season, remain adaptable with your care routine. If you detect yellowing or falling leaves, it might well signal an opportune moment to tweak your regime of watering or light exposure.

The essence of winterizing your container garden lies in harmonizing with the season's natural rhythms while ensuring your plants are primed to flourish come spring. Through actions such as insulating pots, adjusting indoor light and humidity levels, and meticulously tailoring care routines, you provide the vital support your plants need through their dormant phase.

As we draw this chapter on adapting to climate challenges to a close, remember that gardening is a thriving discourse with nature itself. Every season unfurls unique lessons and opportunities for both your plants and yourself as a gardener to learn and grow. Embrace these changes with an open heart, and you will likely discover joy infused in every phase of your gardening adventure.

CHAPTER 7
SUSTAINABLE AND ECO-FRIENDLY PRACTICES

COMPOSTING FOR CONTAINER GARDENERS: CREATING YOUR OWN FERTILIZER

Picture this: you're working in your kitchen, peeling carrots, and instead of tossing those peels into the trash can—a place where they'd typically contribute to the burgeoning mass of landfill waste—you embark on a transformative journey by turning them into rich, nourishing compost. This simple action revitalizes the very container garden you take such pride in cultivating. Composting transcends mere waste recycling; it is about establishing a sustainable, miniature ecosystem right at your doorstep. By turning what would be kitchen scraps into a valuable resource, you're not only reducing landfill waste but also effectively diminishing the production of methane emissions, which play a significant role as greenhouse gases in climate change. This is a small yet impactful contribution to lessening global environmental challenges. Yet, the process doesn't end with waste reduction. Compost introduces essential organic matter to your soil, improving its structure and augmenting its nutrient content. This transformation results in healthier plants and a more vibrant display of blooms that mark the success of your horticultural efforts.

Starting your composting venture is far less daunting than it may appear, even for those with limited space. A compact compost bin or a sleek tumbler can easily find its home on a balcony or a discreet corner of your yard. To initiate this process, you'll begin by layering green materials such as vegetable scraps and coffee grounds, alongside brown materials like dried leaves and shredded paper. This balance between greens, which are nitrogen providers, and browns, the carbon contributors, is critical for the decomposition process. For novices, aiming for a balanced ratio of 1:2 between greens and browns is advisable for achieving optimal results. The compost pile should maintain a level of moisture akin to a wrung-out sponge—neither too dry nor overly saturated—and it's crucial to aerate it by turning it weekly. This practice accelerates the composting process by introducing vital oxygen, which sustains the microorganisms responsible for decomposition.

An understanding of what can be tossed into your compost pile is just as vital as knowing what to exclude. Vegetable scraps are ideal candidates—imagine a variety of carrot tops, potato peels, and wilted spinach leaves. In addition, coffee grounds, eggshells, and tea bags contribute effectively to the organic mix. However, certain items such as meat and dairy products should be avoided, as they are prone to attracting pests and developing unpleasant odors. Similarly, steer clear of oily foods and anything that has been treated with chemicals, as these can disrupt the compost's natural balance. By adhering to natural materials, you can ensure that your compost remains both nutrient-rich and safe for your beloved plants.

Once your efforts have produced a batch of rich, earthy compost, it's time to apply it within the realm of your container garden. Incorporate the compost into your potting soil to enhance its fertility, thereby providing your plants an advantageous start with accessible nutrients right from the roots. Furthermore, you can deploy the compost as a top dressing by gently sprinkling a thin layer over the soil's surface. This method acts as a slow-release fertilizer, gradually doling out nutrients each time you water your plants, fostering growth without the need for chemical alternatives.

Composting Checklist

- **Setup**: Select an appropriately sized bin or tumbler that suits your space.
- **Layering**: Construct alternating layers of green materials (such as veggie scraps) with brown materials (like dry leaves and shredded paper).
- **Maintenance**: Ensure consistent moisture levels, turning the pile regularly each week to promote aeration.
- **Materials**: Incorporate coffee grounds and other organic materials; avoid including meat, dairy, and other non-compostable items.
- **Usage**: Blend with potting soil for enhanced fertility or apply as a top dressing to nourish plants continually.

Composting transcends traditional eco-friendly practices and emerges as a fulfilling, cyclical process of growth and renewal. By embedding it into your gardening routine, you'll observe that it not only considerably reduces waste but also profoundly enriches the life that blossoms around you. This virtuous cycle turns household leftovers into vibrant blooms and lush greenery within your containers. Each motion of the compost pile isn't merely about fostering plant life, but a continuous, positive contribution to the environment. This collaborative effort between nature and your sustainable gardening becomes a powerful testament to the marvels of nature and your enduring commitment to a greener, more sustainable world. Despite being a modest step on a personal level, it has the power to effect significant differences in cultivating a more eco-conscious world.

UPCYCLING CONTAINERS: CREATIVE AND SUSTAINABLE SOLUTIONS

Imagine transforming your living space with a splash of creativity, breathing new life into items once destined for the landfill. Upcycling is more than just a trend; it's a vibrant expression of sustainability and individuality. By repurposing old materials, you contribute to environmental conservation, reducing waste by giving these forgotten objects a second chance. This practice not only alleviates the pressure on waste management systems but also curtails the demand for new resources, effectively diminishing your carbon footprint. More than just a green initiative, upcycling allows for personal expression, enabling you to craft unique garden aesthetics that reflect your personality. In doing so, you create a garden that stands out, not just for its beauty but also for its commitment to sustainability.

With a bit of imagination, ordinary household items can become extraordinary planters. Consider those old boots gathering dust in your closet. By transforming them into quirky planters, you add character and whimsy to your garden, while also keeping those boots out of the trash. Simply fill them with soil and a plant of your choice, and watch as they become a charming focal point. Similarly, mason jars can find new life as herb gardens on your kitchen windowsill. Their clear glass beautifully showcases the soil layers and root structures, turning a simple jar into an intriguing display. For a rustic touch, hang them with twine or wire to create a suspended herb garden that's both practical and visually appealing.

Before starting any upcycling project, it's important to prepare your materials carefully. This ensures your creations are both functional and durable. When repurposing items like cans or jars, drill small drainage holes at the bottom to prevent water from accumulating and causing root rot. For items made of porous materials, like old wooden crates or terracotta pots, apply a non-toxic sealant inside to prevent leaks and extend their lifespan. These simple steps are crucial for ensuring your plants thrive in their new homes. Not only do they help maintain optimal growing conditions, but they also protect your creations from wear and tear over time.

Upcycling is all about creativity and experimentation. Don't be afraid to mix materials and colors for an eclectic look that's uniquely yours. Combine metal tubs with wooden crates or weave textiles into your designs for added texture and warmth. The possibilities are endless, limited only by your imagination. Share your creations on social media platforms to inspire others and join a community of upcyclers who are passionate about sustainability and innovation. This not only spreads awareness about eco-friendly practices but also connects you with like-minded individuals who share your enthusiasm for creative gardening.

A great way to delve deeper into upcycling is by joining online forums or local workshops dedicated to sustainable gardening practices. These spaces offer a wealth of knowledge and inspiration, helping you refine your skills and discover new ideas. They also provide opportunities to showcase your projects and receive feedback from fellow enthusiasts. As you continue to explore this rewarding practice, remember that upcycling is not just about the end result; it's about the process of transforming something ordinary into something extraordinary. This journey fosters creativity, encourages sustainable living, and ultimately leads to a more personalized and meaningful connection with your garden.

Upcycling is more than a hobby; it's a lifestyle choice that reflects a commitment to sustainability and creativity. It invites you to see potential where others see waste, transforming discarded items into beautiful garden pieces that tell a story. Each upcycled container becomes a testament to innovation and resourcefulness, offering a unique blend of function and art. As you experiment with different materials and designs, you'll find that upcycling enriches not only your garden but also your life, fostering a deeper appreciation for the beauty in repurposing and reinvention.

So why not take a look around your home or neighborhood for items that can be given a second life? Let your imagination run wild as you transform everyday objects into stunning garden features that reflect your commitment to both style and sustainability. In doing so, you'll create spaces that are not only environmentally friendly but also uniquely yours—a reflection of your creativity, values, and passion for gardening in harmony with nature.

WATER-WISE GARDENING: REDUCING YOUR ENVIRONMENTAL IMPACT

Water-wise gardening is all about efficiency. It's about choosing plants that naturally thrive with less water and optimizing how and when you water them. Drought-tolerant plants are your best friends here. Think succulents, lavender, or yarrow. These hardy species have adapted to survive with little moisture. They store water in their leaves or have deep root systems that tap into groundwater. By selecting these resilient plants, you create a garden that can withstand dry spells without constant attention.

Once you've chosen your plants, consider how you will water them. Timing is everything. Watering early in the morning or late in the evening reduces evaporation, ensuring more moisture reaches the roots. It's like giving your plants a refreshing drink just when they need it most. This timing also helps prevent diseases that can thrive on wet foliage left to sit overnight.

Installing a drip irrigation system is another effective strategy. This system delivers water directly to the plant's roots, minimizing waste and maximizing efficiency. It's like having a personal assistant for your garden, making sure each plant gets exactly what it needs without the guesswork. Drip systems are easy to set up and can be adjusted for different plant types, ensuring a customized watering plan.

Capturing rainwater is another excellent way to conserve resources. Installing a rain barrel under your downspout collects free water from the sky, which you can use to hydrate your garden during dry periods. This not only saves on your water bill but also makes use of natural resources that might otherwise be wasted. Rainwater is free from the chemicals found in tap water, making it a healthier option for your plants.

Mulching plays a significant role in water conservation, too. A layer of mulch acts like a protective blanket for your soil. Organic options like straw or leaf litter are ideal, as they gradually decompose and enrich the soil while retaining moisture. Mulching keeps the soil cool and reduces evaporation, ensuring your plants stay hydrated longer. Plus, it suppresses weeds, which would otherwise compete with your plants for precious water.

Different plants benefit from various mulching techniques. For instance, vegetables might do well with straw mulch, which is light and easy to move around as needed. Ornamental plants might enjoy leaf litter or bark mulch for a more aesthetic touch. Experiment with what works best for your garden, keeping in mind the specific needs of each plant species.

Engaging with community initiatives on water conservation can amplify your efforts. Local workshops often provide valuable insights into sustainable gardening practices tailored to your region's climate and resources. Participating in these events not only expands your knowledge but also connects you with fellow gardeners who share your passion for sustainability.

Advocating for sustainable practices within community gardens can inspire broader change. By sharing your success stories with neighbors or collaborating on community projects like setting up rainwater collection systems, you contribute to a collective effort towards environmental responsibility. These actions foster a sense of community and show how individual efforts can lead to significant impacts.

Gardening with a focus on water conservation transforms your space into an environmentally conscious haven. The practices not only support plant health but also align with broader goals of sustainability and resource management. As you implement these strategies, you'll find a deeper connection to your garden and the natural world around you.

Creating a water-wise garden is an ongoing process of learning and adaptation. Each season presents new challenges and opportunities for growth, both for you and your plants. Embrace the journey of discovering what works best in your unique environment, knowing that every small step you take adds up to meaningful change. With each droplet saved, each plant that flourishes under your care, you contribute to a larger narrative of sustainability and stewardship of our planet's precious resources.

Water-wise gardening isn't just about conserving water; it's about cultivating mindfulness and intentionality in how we interact with our environment. It's about recognizing our role in the ecosystem and making choices that reflect our values of stewardship and care for the earth. As we continue to explore sustainable practices in our gardens, we find not only beauty and abundance but also a profound sense of purpose and connection to the natural world.

ORGANIC GARDENING PRACTICES: SAFE AND NATURAL TECHNIQUES

Organic gardening is more than just avoiding synthetic chemicals; it's a commitment to fostering a harmonious relationship with nature. This approach emphasizes biodiversity, building healthy soil, and nurturing plants without resorting to artificial aids. By steering clear of synthetic fertilizers and pesticides, you protect the environment and create a safer space for children and pets. This method promotes a thriving ecosystem right in your backyard or balcony, encouraging beneficial insects and microorganisms to flourish.

Managing pests and diseases organically might seem challenging at first, but nature provides its own solutions. Homemade insecticidal soap sprays are a practical option. You can make these by mixing mild soap with water, which effectively tackles soft-bodied insects like aphids without harming beneficials like ladybugs. Another powerful ally is companion planting. Certain plants deter pests naturally or attract beneficial insects that prey on garden nuisances. For instance, marigolds repel nematodes, while basil keeps mosquitoes at bay. Placing these plants strategically can significantly reduce pest issues, making your garden healthier without the need for chemicals.

Soil health is the backbone of any flourishing garden. Organic amendments like worm castings and rock dust play a pivotal role here. Worm castings are essentially worm manure, packed with nutrients that improve soil fertility and boost plant growth. They're easy to incorporate into your garden routine—just mix them into the top layer of soil or use them as a side dressing around your plants. Another treasure for your soil is rock dust. Rich in trace minerals, it enhances soil structure and helps plants absorb nutrients more efficiently. Regularly adding these amendments ensures your soil remains healthy, supporting robust plant growth.

Sourcing organic seeds and plants is another cornerstone of sustainable gardening. Choosing organic options means they haven't been treated with harmful chemicals, supporting cleaner, safer food production. Heirloom varieties are particularly valuable. These seeds pass down unique traits through generations, helping preserve genetic diversity—a vital component in resilient ecosystems. By planting heirlooms, you're not only growing food but also contributing to biodiversity. Supporting local organic seed suppliers further strengthens community ties and promotes sustainable agriculture.

Switching to organic gardening might seem daunting initially, but it's about making small changes that collectively make a big difference. Start by gradually replacing synthetic products with organic alternatives. Observe how your garden responds and adjust accordingly. You'll soon discover that organic gardening fosters a deeper connection to the natural world, encouraging you to learn from the environment around you.

Embracing organic practices transforms your garden into a vibrant ecosystem where every element plays its part. It's about working with nature rather than against it, creating a balanced environment where plants thrive with minimal intervention. As you cultivate your organic garden, you'll find that this approach not only benefits your plants but also enriches your understanding of the intricate web of life that surrounds us.

Engaging in organic gardening also opens the door to community involvement. Joining local gardening groups or attending workshops can provide valuable insights and support as you navigate this journey. Sharing experiences with fellow gardeners helps build a network of like-minded individuals committed to sustainable practices.

As you explore organic gardening, remember that it's a continuous process of learning and adaptation. Each season presents new challenges and opportunities for growth, both for you and your plants. Embrace these experiences as part of the journey towards creating a healthier, more sustainable garden.

The joy of organic gardening lies in its simplicity and effectiveness. It's about finding creative solutions to challenges and discovering the satisfaction that comes from growing plants naturally. Whether you're cultivating vegetables, herbs, or flowers, organic practices offer a holistic approach that nurtures both your garden and your soul.

Incorporating these techniques into your gardening routine transforms it from a mere hobby into an enriching experience that supports biodiversity and promotes environmental stewardship.

As you continue on this path, you'll discover that organic gardening is not just a method but a way of life—a meaningful connection to the earth and its rhythms.

BUILDING A SUSTAINABLE CONTAINER GARDEN: LONG-TERM TIPS

Creating a sustainable container garden isn't just about what you plant today but how you nurture it over time. One key practice is rotating your crops to prevent soil depletion. Just like in traditional farming, different plants use and replenish various nutrients. By changing what you grow in each pot seasonally, you keep the soil balanced and healthy. Imagine your basil thriving this summer, only to swap places with a lovely array of radishes come fall. This simple rotation helps break cycles of pests and diseases, ensuring a fresh start for each planting season.

Encouraging beneficial insects is another natural way to enhance your garden's health. These tiny allies, like ladybugs and bees, play a crucial role in pollination and pest control. You can attract them by planting flowers like marigolds or lavender alongside your veggies. This biodiversity creates a balanced ecosystem where plants thrive, and harmful pests stay at bay. It's like inviting helpful little friends to your garden party, each bringing something valuable to the table.

Minimizing waste in your garden is not only eco-friendly but also cost-effective. Consider using biodegradable pots for seedlings. They break down naturally in the soil, adding nutrients as they decompose. This eliminates the need for plastic containers and reduces waste. When your plants outgrow their pots, simply plant them directly into larger containers or the ground. Composting plant debris is another excellent way to recycle nutrients back into your garden. Leaves, spent flowers, and pruned branches can all be composted rather than discarded. This reduces waste and enriches the soil, closing the loop on your gardening cycle.

Maintaining garden health requires regular attention and care. Begin with soil testing to understand its nutrient profile, making amendments as needed. Kits are available that make this process straightforward, helping you identify deficiencies before they affect plant health. Adding organic matter like compost improves soil structure and fertility, giving your plants the best possible environment to grow. Pruning and grooming are equally important for plant vitality. Regularly remove dead or diseased leaves and trim overgrown branches to promote airflow and light penetration. This not only keeps plants looking neat but also prevents disease.

Continuous learning is vital in sustainable gardening. New techniques and practices emerge regularly, offering innovative ways to improve your garden's sustainability. Reading books, articles, or blogs keeps you informed about the latest trends and ideas. Attending workshops or online courses provides hands-on experience and connects you with other gardeners who share your passion for sustainability. Engaging with community groups or online forums can also be a great source of inspiration and support, especially when facing challenges.

Experimentation is an integral part of sustainable gardening. Trying out new plants or techniques can lead to exciting discoveries about what works best in your space. Keep a garden journal to track your successes and failures, noting what you learn along the way. This record becomes invaluable over time, providing insights that help refine your approach each season. Sharing your experiences with others fosters a sense of community and encourages collaboration on sustainable gardening projects.

Gardening sustainably in containers is a dynamic process that evolves with each season and every new idea you explore. It's about creating a living space that reflects both your values and creativity while contributing positively to the environment. By implementing these long-term strategies, you cultivate not just plants but a thriving ecosystem that supports biodiversity and resilience.

Every small change you make today lays the groundwork for a healthier garden tomorrow, so don't hesitate to experiment and adapt as you go along.

SUPPORTING LOCAL WILDLIFE: CREATING HABITATS IN SMALL SPACES

Imagine the unyielding joy that fills your spirit as you wander through your garden—the delightful array of colors, the intoxicating aromas, and the lively presence of countless living organisms. Picture adding to this the soft, engaging hum of bees darting from blossom to blossom or the delicate, mesmerizing dance of butterflies flitting around. By nurturing a garden that bolsters local wildlife, you're not merely offering a lifeline to various creatures; you are enriching the whole ecological network that is interconnected with your own charming oasis of green. Pollinators, such as bees and butterflies, are integral in ensuring the flora in your garden efficiently produce their extraordinary blooms and nourishing fruits. Meanwhile, birds, along with beneficial insects, are nature's soldiers, tirelessly working to regulate pest populations, maintaining a natural balance. Hence, a garden teeming with diversity becomes an emblem of health and vibrancy.

Creating an inviting environment for a myriad of wildlife in your smaller garden space need not be a daunting task. The journey can simply begin with the introduction of native flowers, which act as a beckoning call to local pollinators. These plants, with their natural evolution alongside indigenous fauna, provide the perfect nectar and pollen that pollinators have long come to rely upon. Think of it as laying out a sumptuous spread brimming with their all-time favorite delicacies. Plants like coneflowers, black-eyed Susans, and milkweed stand as stellar selections to beckon bees and butterflies. Beyond flora, adding a humble bird feeder or even a simple basin of water can work wonders in drawing avian visitors. These birds become more than just intriguing sights to behold; they are active participants in pest control, consuming insects that would otherwise be detrimental to your delicate plants.

Venturing further into supporting various species, try creating miniaturized habitats within your garden containers. Introducing a small log or strategically scattered stones can serve as sanctuary zones for beneficial insects such as beetles and ladybugs. These industrious creatures play essential roles in breaking down organic materials and naturally managing plant pests. Consider also including a tiny water feature — even a shallow dish filled with fresh water can attract amphibians like frogs or offer a much-needed sip to birds and insects. Each of these elements weaves together a small but profound ecosystem within your garden, nurturing and sustaining life across various levels.

Broader community involvement can significantly amplify your efforts to foster local wildlife. By joining or initiating local wildlife conservation groups, you connect with others who share your passion for preserving nature. Working together on community initiatives such as habitat restoration projects not only bolsters environmental well-being but also creates bonds of friendship and shared purpose among participants. Whether it's planting native species along a scenic local trail or fashioning wildlife-friendly zones in urban areas, every little action contributes to a sizable environmental impact.

When you inspire friends and neighbors to adopt wildlife-inclusive practices, you expand your influence beyond the confines of your own garden. Sharing seeds, offering gardening tips, and exchanging success anecdotes serves to motivate others, collectively forming a network of gardens that enhance biodiversity. Such synergy acts as a ripple, extending positive outcomes far beyond personal efforts, fortifying local ecosystems as more gardens transform into refuges for wildlife.

By aiding wildlife, you embed yourself into the sweeping story of conservation and stewardship. Your garden, while still a personal escape, becomes an essential piece in the larger ecological puzzle, contributing to the fortitude and flourishing of local ecosystems. The vibrant sights and harmonious sounds brought by wildlife infuse an added depth of delight into your gardening pursuits, making it a living testament to the profound and beautiful interconnectedness of nature.

In this chapter, we've unfolded how even compact spaces can yield significant environmental contributions through sustainable endeavors like composting, upcycling, judicious water management, organic gardening, and creating hospitable environments for wildlife. Each tactic not only enriches your surroundings but also uplifts environmental health. As we advance to the next chapter, we'll explore ways to tackle common challenges in container gardening, equipping you with tangible solutions to secure ongoing success with your blooming, eco-friendly garden.

These methodologies do more than just benefit the environment; they metamorphose your gardening routine into a pursuit that's deeply meaningful and impactful. With every sustainable step, you're nurturing the planet, gradually cultivating it into a healthier version of itself, container by container. Embrace these practices wholeheartedly and relish in observing your garden's transformation into a thriving, self-sustaining ecosystem.

CHAPTER 8
TROUBLESHOOTING AND SUCCESS STORIES

COMMON CONTAINER GARDENING MISTAKES AND HOW TO AVOID THEM

One of the most frequent mistakes is overcrowding. The urge to cram many plants into a single pot can be overwhelming. Initially, it might seem like a good idea, as you imagine a pot brimming with life, with the plants standing tall like a supportive community. However, the reality is that they will compete intensely for the limited nutrients and water available, which ultimately stunts their growth. A better approach is to give each plant the breathing room it deserves. A good rule of thumb is to limit your selection to one or perhaps two plants for larger containers. This ensures that each has enough room to spread its roots and access the resources it needs to thrive beautifully.

Additionally, pitfalls lie in the choice of soil. Garden soil from your backyard may appear handy and cost-effective, but it compacts too easily, which can suffocate roots and prevent adequate drainage. This can be a silent saboteur in your gardening efforts. Instead, opting for container-specific potting mixes is highly recommended. These specialized blends are scientifically formulated to provide the optimal balance of nutrients and aeration needed for container plants. Common ingredients include peat moss or coconut coir, which have natural moisture-retaining qualities without risking the dreaded waterlogging of your plants. Choosing appropriate soil is undeniably crucial for promoting healthy root development and ensuring overall plant vigor.

A little research into each plant's unique needs can make a world of difference. Familiarize yourself with their light, water, and soil preferences before your trowel touches earth. Planning your garden layout thoughtfully can prevent potential problems. Consider how much space each plant will require as it grows, and evaluate how much sunlight your garden area receives throughout the day. By sketching a rough plan before planting, you'll ensure that your garden is not only aesthetically pleasing but also functional and sustainable over time.

The impacts of these initial mistakes can be significant. Overcrowded pots result in reduced yields as plants vie for the limited resources they so desperately need. This competition also makes them more vulnerable to pests and diseases, as stressed plants can attract unwanted visitors looking to exploit their weakened state. Inadequate soil leads to poor drainage, which, in turn, can cause root rot—a stealthy killer in many container gardens. By addressing these concerns proactively, you can save yourself both time and frustration. This preparation fosters a bountiful harvest of healthy, thriving plants, transforming what could have been a disappointing venture into a rewarding and visually spectacular outcome.

INTERACTIVE EXERCISE: PLANNING YOUR PERFECT CONTAINER GARDEN

Take a creative journey by grabbing a notebook or even a whiteboard, and let your imagination run wild by sketching your ideal container garden layout. As you do, note the sunlight exposure for each potential spot, and strategically plan which plants will inhabit each area based on their specific light needs and mature size. Armed with this visual blueprint, be sure to list any questions or uncertainties you may have about the specific requirements of certain plants. To gather the answers, delve into reputable online gardening resources or consult traditional gardening guides. This exercise is more than just a plan; it's a powerful visualization of your garden's eventual potential, and it serves as a guide for making methodical and informed planting choices.

Mistakes are an inevitable part of the gardening learning curve. They're not seen merely as setbacks, but as opportunities to grow your skills and deepen your relationship with the plant world. By gaining an understanding of these common pitfalls and taking proactive steps to mitigate them, you can cultivate a thriving container garden that consistently brings you joy and satisfaction, season after abundant season. Gardener, embrace the process—from the first seed to the last flower—and watch as you develop not only a beautiful garden but also a deeper understanding and appreciation for nature's delicate balance.

DIAGNOSING PLANT PROBLEMS: SYMPTOMS AND SOLUTIONS

Imagine walking out to your container garden, only to notice that the once vibrant leaves of your beloved basil are turning yellow. It's easy to feel a wave of panic, but fear not—this is where a little detective work comes in handy. Yellowing leaves often signal a nutrient deficiency, specifically nitrogen, which is crucial for lush, green foliage. When you see this symptom, consider adding an organic fertilizer rich in nitrogen to rejuvenate your plant's vigor. Wilting, on the other hand, could be a sign of water imbalance. If your plants look droopy despite regular watering, you might be overwatering them. Check the soil moisture; if it feels soggy, it's time to let it dry out a bit before the next watering session.

To effectively troubleshoot plant problems, having a checklist can be your best ally. Start by examining your plants for signs of pests or diseases. Look under leaves and around stems for any unwanted guests like aphids or fungus gnats. Next, test the soil's pH and moisture levels. A simple pH meter can help you determine if the soil is too acidic or alkaline, which can impact nutrient availability. For moisture, stick your finger into the soil about an inch deep—if it feels dry, your plant might need a drink. Alternatively, if it feels too wet, scaling back on watering is wise. Observing these factors can often lead you directly to the root of the issue.

Understanding what causes plant problems is key to prevention and recovery. Poor drainage is a leading cause of root rot, a condition where roots become waterlogged and begin to decay. Ensure your containers have adequate drainage holes and avoid using heavy garden soil that retains too much water. Similarly, insufficient sunlight can stunt plant growth, leaving them spindly and weak. Make sure your sun-loving plants get at least six hours of direct sunlight daily to thrive. If lighting is an issue, consider rearranging your garden to optimize sun exposure or investing in grow lights for indoor setups.

Once you've identified the problem, implementing effective remedies is crucial. Adjust watering schedules based on your plant's specific needs—most prefer consistent moisture but not constant saturation. When nutrient deficiencies are the culprit, incorporate organic fertilizers or soil amendments to replenish missing elements. Products like worm castings or compost can work wonders in providing a balanced nutrient profile. These solutions not only restore health but also strengthen your plants against future stressors.

CONTAINER GARDEN TROUBLESHOOTING CHECKLIST

Symptom	Possible Cause(s)	Steps to Diagnose & Resolve
Yellowing Leaves	- Nitrogen deficiency - Overwatering or poor drainage - Root-bound plant	✅ Check for soggy soil and drainage holes ✅ Test soil nitrogen levels or fertilize with balanced fertilizer ✅ Gently remove plant from pot to check for root crowding
Wilting (even with wet soil)	- Root rot from overwatering - Poor drainage or compacted soil - Fungal disease	✅ Inspect roots for dark, mushy texture (sign of rot) ✅ Improve drainage or repot in fresh, well-aerated mix ✅ Disinfect pot and prune dead roots
Wilting (dry soil)	- Underwatering - Too small a container - Excessive sun or wind	✅ Water deeply and slowly ✅ Upgrade to a larger pot if roots are bound ✅ Provide shade during peak heat
Brown Leaf Tips or Edges	- Salt buildup from fertilizer - Low humidity or underwatering - Pot too hot in direct sun	✅ Flush pot with clean water to remove excess salts ✅ Mist leaves or increase humidity ✅ Place pots on trays or use insulating materials

Symptom	Possible Cause(s)	Steps to Diagnose & Resolve
Stunted Growth	- Nutrient deficiency (e.g., phosphorus or potassium) - Compact soil or rootbound - Low light levels	☑ Fertilize with a complete slow-release or liquid fertilizer ☑ Check for root crowding and loosen or repot ☑ Relocate to a brighter area
Leaf Spots or Blotches	- Fungal or bacterial disease - Pest damage - Sunburn from water droplets	☑ Remove affected leaves ☑ Apply organic fungicide if needed ☑ Water at soil level and avoid leaf splash
Holes or Chewed Leaves	- Aphids, caterpillars, slugs, beetles	☑ Inspect underside of leaves ☑ Spray with insecticidal soap or neem oil ☑ Handpick large pests or set up traps

Symptom	Possible Cause(s)	Steps to Diagnose & Resolve
Powdery or Sticky Residue	- Powdery mildew - Aphids or whiteflies (honeydew)	☑ Use horticultural oil or fungicide for mildew ☑ Wipe leaves, rinse plants, or spray insecticidal soap
Pale or Discolored Leaves	- Iron or magnesium deficiency - Alkaline soil pH	☑ Use a soil pH test kit (target pH ~6.0–6.8) ☑ Apply chelated iron or magnesium as needed ☑ Consider repotting with acidic potting mix if needed
No Flowers or Fruit	- Too much nitrogen - Not enough light - Wrong temperature range	☑ Use low-nitrogen, high-phosphorus fertilizer ☑ Ensure 6–8 hrs of sunlight/day ☑ Protect plants from temperature extremes

TESTING & MONITORING TIPS

- **Soil Moisture:** Use your finger or a moisture meter to check 2" below the surface.

  - **Soil pH:** Use a pH test kit or digital tester; amend with sulfur (to lower) or lime (to raise).

  - **Nutrient Deficiency:** Look for pattern—older vs. younger leaves affected can point to different deficiencies.

  - **Pests:** Inspect both sides of leaves, stems, and soil surface regularly.

Navigating plant problems might seem daunting at first, but with practice and patience, it becomes second nature. Each symptom tells a story about what's happening below the surface, and learning to interpret these signs enhances your gardening prowess. The satisfaction of nursing a struggling plant back to health is immense—it's like solving a mystery with a happy ending. Keep experimenting and observing; every challenge is an opportunity to deepen your understanding of the fascinating world of container gardening.

SUCCESS STORIES: INSPIRING JOURNEYS FROM BEGINNER TO PRO

In the hustle and bustle of city life, a rooftop garden can be a slice of paradise. Take Emma, a city dweller who transformed her cramped balcony into a lush green haven. When she first started, Emma didn't know the difference between a trowel and a spade. Yet, with dedication, she now grows everything from juicy tomatoes to fragrant herbs in her small space. Her secret? Consistent observation and adaptability. She checks her plants daily, adjusting water and sunlight as needed. This hands-on approach allows her to spot issues early and find creative solutions.

Another inspiring tale is of Jake, who sought self-sufficiency through homegrown veggies. Living in an apartment with no yard, he turned to container gardening as a way to produce his own food. With limited space, Jake became innovative. He used vertical gardening techniques to maximize his growing area, hanging pots on walls to save room. His home is now filled with a bounty of produce, from crunchy bell peppers to sweet strawberries. His success lies in his willingness to learn and experiment, trying new methods until he found what worked best.

Challenges are part of every gardener's story. Emma faced climate issues, with her plants wilting under the harsh summer sun. Rather than give up, she researched shade cloths and set up a makeshift shelter to protect her plants. Jake battled pests early on, losing his first batch of basil to hungry aphids. Instead of being discouraged, he explored natural pest control methods like neem oil, which helped him safeguard future crops. These stories illustrate that setbacks are not failures but stepping stones towards success.

Setting achievable goals is key. Start with something small, like growing a single type of herb or flower, and build from there. Celebrate each milestone—your first flower bloom or the first tomato you pick. These small victories build confidence and inspire further growth. Emma plans to expand her garden by adding more flowering plants next season, while Jake has his sights set on trying his hand at container-grown potatoes. Each goal reached fuels the next adventure.

In the world of container gardening, creativity reigns supreme. Emma uses recycled containers and old buckets as planters, finding beauty in repurposing everyday items. Jake shares his space with local wildlife, setting up a small birdbath that attracts pollinators and adds life to his garden. Their stories prove that you don't need a large plot of land to create something extraordinary. With a bit of imagination and resourcefulness, even the smallest space can become a thriving garden.

Remember that every gardener started as a beginner. The path from novice to expert is paved with trial and error, learning from each experience along the way. Emma and Jake's journeys show that passion and perseverance are the true ingredients for success. So whether you're nurturing a single pot on your windowsill or cultivating a full balcony oasis, embrace the process with enthusiasm.

As you embark on your own gardening adventure, keep in mind that the possibilities are endless. Let Emma and Jake's stories inspire you to try new things and push boundaries. Set your goals high, but take small steps towards them. Celebrate your achievements and learn from your mistakes. Your container garden is not just about growing plants; it's about cultivating joy, creativity, and resilience in your life.

Consider joining local gardening groups or online communities for support and inspiration. Sharing experiences with others can provide valuable insights and encouragement. Swap tips, trade seeds, or simply enjoy the camaraderie of fellow gardeners cheering each other on.

Remember that gardening is an ongoing journey of discovery and growth—a lifelong pursuit where the rewards extend far beyond the harvest itself. Each season brings new opportunities to learn, adapt, and thrive alongside your plants. So grab your trowel, roll up your sleeves, and get ready to create something beautiful— one container at a time.

LEARNING FROM FAILURES: TURNING SETBACKS INTO GROWTH OPPORTUNITIES

Gardening, much like life, is a series of experiments. You plant seeds, nurture them, and sometimes, despite your best efforts, things don't go as planned. It's crucial to remember that setbacks are a natural part of the gardening journey. Plants may fall prey to pests or diseases, regardless of your vigilance. Maybe you chose a plant that just wasn't suited for your climate, and it struggled to survive. These moments can be disheartening, yet they are also rich with lessons waiting to be unearthed.

Reflecting on these experiences is key to growing as a gardener. Keeping a journal is an invaluable tool for this process. Document what you plant, when you plant it, and how it fares over time. Note any issues that arise and how you attempt to resolve them. This practice not only helps track progress but also creates a personal reference guide for future gardening endeavors. Reflecting on what worked and what didn't allows you to make informed decisions moving forward.

When faced with setbacks, resilience is your greatest ally. Instead of seeing failures as roadblocks, view them as opportunities for innovation. Perhaps your tomatoes didn't thrive in the pots you chose; this is a chance to research alternative varieties better suited for containers or your specific climate. Adjusting care routines based on past experiences is crucial. If a particular watering schedule didn't work before, tweak it until you find the sweet spot that keeps your plants happy and healthy.

Stories of resilience abound in the gardening community. Consider Maria, who faced a season of nutrient deficiencies that left her plants looking lackluster. She could have thrown in the towel, but chose instead to revamp her soil management practices. By incorporating organic compost and performing regular soil tests, she transformed her garden into a lush oasis by the next season. Another gardener, Paul, saw his first attempt at growing cucumbers end in failure due to early frost damage. Rather than give up, he studied methods for extending the growing season and successfully replanted, resulting in a bountiful harvest the following year.

Every setback carries the seed of a lesson. When pests invade, it might be time to explore natural pest control methods or companion planting strategies that deter those unwanted guests. If a plant isn't thriving, investigate its specific needs—perhaps it craves more sun or prefers a different soil pH. These explorations deepen your understanding and make you a more adept gardener over time.

The beauty of gardening lies in its endless cycle of growth and renewal. Each season presents new challenges and triumphs, teaching patience and perseverance along the way. Embrace this ongoing learning process with an open heart and a curious mind. You'll find that with each mistake comes the opportunity for creativity and discovery—a chance to reinvent your garden with newfound wisdom.

Setbacks shouldn't deter your enthusiasm for gardening; rather, they should fuel your determination to improve. As you gain experience, you'll learn to anticipate potential issues before they become significant problems. Your confidence will grow alongside your plants, leading to increasingly successful seasons.

Remember that every gardener has faced failures. It's part of what makes gardening such a rewarding pursuit—overcoming challenges and witnessing your hard work come to fruition. Whether it's battling pesky aphids or coaxing reluctant seeds to sprout, each step of the process contributes to your growth as a gardener.

Your garden is a reflection of your resilience and adaptability. Celebrate each small victory—a sprouting seedling or a blooming flower—as proof of your dedication and passion. Let these moments inspire you to continue nurturing your space with love and determination.

Your journey through gardening's ups and downs will shape you into a resourceful and knowledgeable gardener capable of tackling any challenge thrown your way. Embrace the failures as part of the process; they are stepping stones on the path to success in the world of container gardening.

COMMUNITY CONNECTIONS: JOINING THE CONTAINER GARDENING MOVEMENT

Imagine the joy of walking into a room filled with people who share your passion for growing things, eagerly swapping stories about their latest gardening triumphs and challenges. Joining a community of fellow container gardeners can be a game-changer. It opens the door to a wealth of shared knowledge and resources that can transform your gardening experience. When you connect with others, you tap into a collective brain trust, gaining access to tips and techniques that might take years to learn on your own. This network is invaluable, offering advice and support when you're stumped by a stubborn plant problem or need a fresh idea for your next project.

Building friendships within a gardening community doesn't just enrich your gardening life; it also enriches your personal life. These connections often extend beyond the garden, creating lasting relationships and support networks. Whether you're swapping seeds or organizing a group trip to the local nursery, the camaraderie of a gardening community is something special. You'll find yourself not only growing plants but also growing friendships that are rooted in shared experiences and a mutual love for the earth.

Finding these communities is easier than you might think. Local gardening clubs or meetups are fantastic places to start. These groups often host regular meetings where you can learn from more experienced gardeners and share your own insights. If in-person meetings aren't your style, online forums and social media groups offer an alternative. Platforms like Facebook and Reddit have thriving gardening communities where you can ask questions, share photos of your progress, and get advice from people all around the world.

The beauty of these communities lies in their collaborative spirit. Organizing plant swaps or seed exchanges is a popular activity that not only diversifies your garden but also fosters a sense of community. Imagine trading a few of your thriving basil seedlings for a neighbor's heirloom tomato plants—both of you walk away enriched by the exchange. Community garden projects take this collaboration to another level, allowing you to work alongside others to create something beautiful that benefits everyone involved.

The impact of community involvement on personal growth cannot be overstated. When you're surrounded by supportive peers, learning new techniques becomes a shared adventure rather than a solitary task. Seasoned gardeners often take newcomers under their wing, offering guidance that boosts confidence and accelerates learning. Seeing what others have achieved in their own gardens is incredibly motivating; it spurs you on to try new things and push your own boundaries.

Witnessing community achievements firsthand can ignite a passion for experimentation and innovation in your own garden. You might find yourself inspired to try growing an unusual vegetable variety simply because someone in your group has done so successfully. Or perhaps you'll adopt a new method for improving soil health after hearing about the positive results it brought to another gardener's plot.

Participating in these communities also provides an opportunity to give back by sharing your own knowledge and experiences. As you grow in confidence and skill, you'll find yourself offering tips to newcomers who are just starting out. This exchange of ideas and experiences strengthens the community as a whole, creating an environment where everyone learns and grows together.

The role of community in personal growth extends beyond just acquiring new skills; it fosters a sense of belonging and purpose. When you're part of something larger than yourself, your successes feel even more rewarding because they're shared with others who understand the journey. The encouragement and recognition from fellow gardeners boost morale and inspire further exploration.

As you become more involved in gardening communities, you'll likely discover areas where you can contribute uniquely. Perhaps you have a knack for organizing events or an eye for design that helps others plan their garden layouts. Sharing these talents enhances the community while allowing you to develop skills that go beyond gardening.

In essence, joining the container gardening movement is about more than cultivating plants; it's about cultivating connections, creativity, and community spirit. Whether you're attending your first meetup or logging into an online forum for tips on pest control, remember that every interaction is an opportunity to grow —not just as a gardener but as part of a vibrant, supportive network dedicated to making the world a greener place.

PLANNING FOR THE FUTURE: EXPANDING YOUR GARDENING HORIZONS

Picture yourself standing on your balcony or patio, immersing yourself in the lush beauty of your thriving container garden. It's amazing how such a small space, full of potential, can transform into a spectacular oasis of greenery and vitality. Envision a future where your garden becomes not only a source of aesthetic pleasure and daily relaxation but also a bountiful supplier of fresh, home-grown produce. Planning and setting realistic goals for your garden can be both invigorating and motivating. As you ponder your gardening ambitions, consider what you wish to achieve. Could it be an expanded herb collection to elevate your culinary creations, an enhanced assortment of vegetables to yield a plentiful harvest throughout the seasons, or an integration of flowering plants that provide a painterly splash of color year-round? Whatever your unique vision may encompass, formulating clear and achievable goals will effectively chart the course for your gardening journey ahead.

Expanding your garden need not be a rapid endeavor; in fact, gradual and measured growth often yields the most satisfying and sustainable results. Start by methodically adding new containers to your space. This incremental approach allows you to carefully manage the additional responsibilities, ensuring that each new plant receives the individualized attention and care it deserves. By introducing new varieties slowly, you grant yourself the necessary time to thoroughly acquaint yourself with the specific demands and attributes of each plant. This considered approach makes calibrating your care routines more intuitive and seamless. Slowly broadening your garden's diversity will create a vibrant, interconnected tapestry of plants that thrive symbiotically, enhancing both productivity and aesthetic value.

For those poised to delve into more sophisticated techniques, the world of gardening is replete with splendid opportunities and novel methodologies. Imagine venturing into the realms of hydroponics or aquaponics—innovative systems that allow plant cultivation without the traditional reliance on soil, instead utilizing nutrient-rich water. These cutting-edge techniques are especially appealing for those facing spatial limitations, as they optimize profitability and yield. Incorporating smart technology within your garden can revolutionize your approach to nurturing plants. Consider employing sensors or apps that meticulously monitor crucial elements like soil moisture, ambient temperature, and light exposure. The integration of technology can significantly optimize plant health, rendering garden management more streamlined and effective.

As you continue to cultivate and enrich your container garden, adopt a mindset of lifelong learning and adaptability. The field of gardening is one of constant evolution, with exciting trends and innovations emerging at a steady pace. Immerse yourself in advanced gardening literature or consider enrolling in specialized courses to deepen your understanding and inspire novel ideas. Staying informed about groundbreaking practices not only benefits your garden but also maintains your enthusiasm and fulfillment as a gardener. Adaptability is essential to flourishing within the gardening community; embracing change and innovation infuses your garden with vitality and resilience.

Planning for the future of your garden transcends mere physical expansion; it's fundamentally about nurturing a mindset of growth, exploration, and curiosity. As you articulate your goals and experiment with new methodologies, remember that gardening is a voyage of discovery, teeming with opportunities to learn and develop in harmony with your plants. Whether you're dreaming of harvesting vine-ripened tomatoes or crafting a sanctuary that invites the comforting buzz of pollinators, each deliberate step taken brings you closer to realizing your gardening dreams.

In conclusion, envisioning the future of your container garden involves dreaming boldly while undertaking small, deliberate steps toward realizing those dreams. Your garden is not just a collection of plants; it's a living canvas for creativity, a space that invites you to experiment with ideas and embrace the potential of what a seemingly modest space can offer. Through mindful planning and a harmonious blend of technology and tradition, you can cultivate a garden that genuinely resonates with your personal style and values, meeting both your practical needs and aesthetic desires.

KEEP THE GARDEN GROWING

Now that you've learned how to grow your own veggies, herbs, and flowers in containers, you've got everything you need to turn even the smallest space into a garden.

But here's something else you can grow—**a little encouragement** for the next person just getting started.

By sharing your honest thoughts about this book on Amazon, you'll help others find the same tips and confidence that helped you. It's like leaving a trail of sunshine for the next gardener to follow.

Thank you for being part of this growing community. Container gardening keeps blooming when we share what we've learned—and you're helping make that happen.

☞ **Scan the QR code or go here to leave your review on Amazon:** https://www.amazon.com/review/review-your-purchas es/?asin=B0FD478RY5

Thanks again—and happy planting!

– Avery Sage

CONCLUSION

As we wrap up this journey together, let's take a moment to appreciate how far you've come. Just a while ago, the idea of creating your own container garden might have seemed daunting. But look at you now, armed with knowledge and confidence, ready to transform any small space into a lush, vibrant oasis.

This book has walked you through the essentials of container gardening. We started with understanding the basics, like choosing the right containers and creating the perfect soil mix. We explored the art of selecting and caring for plants, whether they are vegetables, herbs, or flowers. We delved into watering techniques, fertilization strategies, and even how to deal with pests and diseases naturally. Each chapter was designed to build your skills step-by-step, making the process approachable and enjoyable.

Key takeaways from our journey include the importance of understanding your plants' needs. From sunlight and water to nutrients and space, every element plays a role in their health. You've learned how to identify and solve common problems, ensuring your plants thrive. And let's not forget the creative aspects —designing your garden layout, experimenting with vertical gardening, and integrating companion planting for a harmonious ecosystem.

Now, I invite you to take action. If you haven't started your container garden yet, now is the time. Use this book as your guide, but don't be afraid to experiment and make it your own. Try new plant combinations, explore different gardening techniques, and most importantly, have fun with it. Your garden is a reflection of your creativity and care, and there's no right or wrong way to grow it.

Looking ahead, envision your future as a gardener. Imagine your space filled with thriving plants, each one a testament to your dedication and growth. Consider how this journey might inspire you to expand your garden, try new techniques, or even share your newfound knowledge with others. The possibilities are endless, and your garden will grow with you as you continue to learn and explore.

Remember, gardening is not just a hobby; it's a lifelong journey. There will be moments of triumph and times of challenge, but each experience adds to your story. Embrace the process, celebrate the small victories, and learn from the setbacks. Your garden is your canvas, and you are the artist.

As we conclude, I want to thank you for choosing to embark on this journey with me. Your commitment to learning and growing is truly inspiring. I hope this book has equipped you with the tools and confidence to create a beautiful, flourishing container garden. May it bring you joy, peace, and a deeper connection to the natural world. Happy gardening!

REFERENCES

1. themicrogardener.com. (2025). *The Benefits of Container Gardening.* https://themicrogardener.com/the-benefits-of-container-gardening/
2. gardenerspath.com. (2025). *Containers, Pots, and Planters: What Material Is Best?* https://gardenerspath.com/how-to/containers/plant-containers-pots-planters-material-best/
3. extension.psu.edu. (2025). *Homemade Potting Media.* https://extension.psu.edu/homemade-potting-media
4. www.farmstandapp.com. (2025). *12 Container Gardening for Urban Spaces Tips That ...* https://www.farmstandapp.com/6011/container-gardening-for-urban-spaces/
5. extension.umd.edu. (2025). *Growing Vegetables in Containers.* https://extension.umd.edu/resource/growing-vegetables-containers
6. gardenary.com. (2025). *How to Grow Lots of Herbs in a Small Space - Gardenary.* https://www.gardenary.com/blog/how-to-grow-herbs-in-a-small-space
7. hgtv.com. (2025). *24 Easy Flowers for Beginners to Grow.* https://www.hgtv.com/outdoors/flowers-and-plants/flowers/13-cant-kill-flowers-for-beginners-pictures
8. permacultureapartment.com. (2025). *Container Garden Companion Planting Guide.* https://www.permacultureapartment.com/post/container-garden-companion-planting
9. rainbird.com. (2025). *How to Plan an Automatic Drip Watering System ...* https://www.rainbird.com/homeowners/blog/how-to-plan-an-automatic-drip-watering-system-for-container-plants
10. earthbox.com. (2025). *Planter Boxes: 10 Benefits for Urban Gardening.* https://earthbox.com/blog/planter-boxes-for-urban-gardening
11. milorganite.com. (2025). *Organic vs Synthetic Fertilizer - Milorganite.* https://www.milorganite.com/lawn-care/organic-lawn-care/organic-vs-synthetic
12. plushbeds.com. (2025). *13 Ways to Implement Water Conservation in Your Garden.* https://www.plushbeds.com/blogs/green-sleep/13-ways-to-implement-water-conservation-in-your-garden
13. rootsandrefuge.com. (2025). *A Complete Guide to Vertical Gardening (On a Budget!)* https://rootsandrefuge.com/vertical-gardening-on-a-budget/

REFERENCES

14. finegardening.com. (2025). *The Elements of Great Garden-Container Design* ... https://www.finegardening.com/article/the-elements-of-great-garden-container-design-simplified

15. vickiodell.com. (2025). *15 Upcycled Container Projects*. https://vickiodell.com/15-upcycled-container-projects/

16. xerces.org. (2025). *Pollinator-Friendly Native Plant Lists*. https://xerces.org/pollinator-conservation/pollinator-friendly-plant-lists

17. gardeningknowhow.com. (2025). *Container Garden Pest Control*. https://www.gardeningknowhow.com/special/containers/container-pests.htm

18. eartheasy.com. (2025). *Natural Garden Pest Control: Safe, Non-Toxic Methods* ... https://learn.eartheasy.com/guides/natural-garden-pest-control/

19. dummies.com. (2025). *Common Plant Diseases in Container Gardens*. https://www.dummies.com/article/home-auto-hobbies/garden-green-living/gardening/containers/common-plant-diseases-in-container-gardens-180895/

20. npic.orst.edu. (2025). *Beneficial Insects - National Pesticide Information Center*. https://npic.orst.edu/envir/beneficial/index.html

21. themicrogardener.com. (2025). *Guide to Understanding Microclimates in your Garden*. https://themicrogardener.com/guide-understanding-microclimates-in-your-garden/

22. planthardiness.ars.usda.gov. (2025). *USDA Plant Hardiness Zone Map*. https://planthardiness.ars.usda.gov/

23. fbfs.com. (2025). *How to Protect Crops From Extreme Weather*. https://www.fbfs.com/learning-center/protect-crops-from-extreme-weather

24. cyclandscaping.com. (2025). *Principles Of Xeriscaping And Its Benefits*. https://cyclandscaping.com/principles-of-xeriscaping-and-its-benefits/

25. piedmontmastergardeners.org. (2025). *Home Composting Solutions For Virtually Everyone*. https://piedmontmastergardeners.org/article/composting-options-for-small-indoor-and-restricted-spaces/

26. bhg.com. (2025). *24 Unique Repurposed Planters Made from Salvaged* ... https://www.bhg.com/gardening/container/plans-ideas/beyond-the-ordinary-flowerpot/

27. ngb.org. (2025). *The Ultimate Guide to Water-Wise Gardening*. https://ngb.org/water-wise-gardening-plants/

28. esteelicious.com. (2025). *Top 5 Natural Pest Control Methods for Container Gardens*. https://www.esteelicious.com/top-5-natural-pest-control-methods-for-container-gardens/

29. provenwinners.com. (2025). *10 Container Gardening Mistakes to Avoid*. https://www.provenwinners.com/learn/top-ten-lists/10-container-gardening-mistakes-avoid

30. pubs.ext.vt.edu. (2025). *Diagnosing Plant Problems | VCE Publications - Virginia Tech*. https://www.pubs.ext.vt.edu/426/426-714/426-714.html

31. homestead.org. (2025). *Container Gardening in the City: Urban Homesteading on ...* https://www.homestead.org/gardening/container-gardening-in-the-city-urban-homesteading-on-a-budget/

32. isminc.com. (2025). *10 Reasons to Join a Community Garden*. https://isminc.com/advisory/publications/the-source/10-reasons-to-join-a-community-garden

THE ULTIMATE BEGINNER'S GUIDE TO RAISING CHICKENS

BACKYARD CHICKEN KEEPING MADE EASY—RAISE HAPPY HENS AND ENJOY FRESH EGGS EVEN IF YOU'RE A TOTAL BEGINNER!

INTRODUCTION

Many years ago, I stayed at a bed and breakfast in upstate New York. One sunny morning, I stood in the backyard, a steaming cup of coffee in hand, watching the sunrise cast a golden glow over the owners' small flock of chickens. They clucked contentedly, scratching and pecking at the ground, completely absorbed in their world. It was a simple moment of joy that captured the essence of why so many people are turning to backyard chicken-keeping. The satisfaction of gathering fresh eggs and the connection to nature it brings are unmatched.

This book aims to be your go-to guide for raising chickens, especially if you are just beginning. My goal is to equip you with all the foundational knowledge you need to start and maintain a healthy and productive flock. Whether you're looking to enjoy fresh eggs, practice sustainable living, or simply find joy in caring for these delightful creatures, this book is here to support you every step of the way.

Raising chickens comes with numerous benefits. There's the obvious reward of fresh eggs, but it goes beyond that. By keeping chickens, you take a step towards sustainable living. You create a small ecosystem in your backyard and connect with a more natural way of life. Plus, chickens have a way of bringing unexpected joy. They have personalities, quirks, and routines that can quickly become a beloved part of your daily life.

The structure of this book is designed to guide you through every stage of chicken-keeping. We'll start with selecting the right breed that suits your needs and environment. Then, we'll cover setting up the coop, ensuring it's safe and comfortable. You'll learn how to nurture chicks into healthy adults and manage their health and wellness. Each chapter builds on the last, providing you with a clear roadmap for your journey.

It's natural to have concerns as a beginner. You might worry about failing or not knowing enough. Rest assured, this book provides clear, step-by-step guidance to build your confidence. You'll find practical advice and tips to handle common challenges. Mistakes will happen, but they are part of the learning process. With patience and persistence, your efforts will be rewarded.

This guide focuses on small-scale, backyard operations. It's tailored for those who seek self-sufficiency and sustainable living. The advice is practical and actionable, designed to fit into your lifestyle and align with your goals.

As you read, I invite you to engage actively with the material. Take notes, reflect on your personal goals, and explore additional resources provided throughout the book. This will deepen your understanding and help you tailor the advice to your unique situation.

In closing, I encourage you to take that first step towards chicken-keeping. Apply what you learn and enjoy the rewarding experience of raising your own flock. Start small, stay curious, and embrace the journey. Before you know it, you'll be gathering fresh eggs and watching your chickens with your own sense of wonder and fulfillment. Welcome to the adventure.

CHAPTER 1
GETTING STARTED WITH BACKYARD CHICKENS

My journey into the world of backyard chickens began at a local farmer's market, witnessing a family's joy as they prepared to welcome their new feathered friends. Their enthusiasm was infectious, highlighting the increasing popularity of raising chickens at home. Beyond the allure of fresh eggs, this endeavor represents a deeper venture into self-reliance, environmental stewardship, and forging a closer bond with the natural world. This chapter delves into the foundational step of this fulfilling hobby: selecting the most suitable chicken breed for your needs.

CHOOSING THE RIGHT CHICKEN BREED FOR BEGINNERS

Selecting the right chicken breed is crucial to your success as a new chicken keeper. It's akin to choosing a pet that fits your lifestyle, yet it's also a commitment to understanding their unique needs and behaviors. Different breeds offer varying benefits, from egg production to temperament and adaptability. For instance, if fresh eggs are your primary goal, consider prolific layers like Rhode Island Reds or Leghorns. These breeds can offer a nearly year-round supply of eggs, providing consistent yields that help reduce trips to the grocery store. Rhode Island Reds are particularly hardy and easy to care for, showing resilience in diverse environments, which makes them ideal for beginners focused on egg production.

When it comes to temperament, having chickens that are friendly and easygoing is imperative, especially if you have children or neighbors close by. Buff Orpingtons and Australorps are known for their calm demeanor, making them excellent ambassadors for the joys of chicken-keeping. Buff Orpingtons, often described as the "golden retrievers" of the chicken world, are gentle, approachable, and dependable in their interactions with both humans and other chickens, making them a delightful addition to any backyard flock.

Climate adaptability is another critical factor. If you live in an area with harsh winters or hot summers, choosing breeds that can withstand temperature extremes will save you headaches down the road. Plymouth Rocks and Wyandottes, for instance, are well-suited for variable climates. Their robust feathering insulates them against the biting cold while still allowing them to remain comfortable and relatively cool during warmer months. These breeds exemplify resilience and adaptability, refining the art of chicken-keeping even in less forgiving climates.

The space you have available and your neighborhood's noise tolerance are also essential considerations. If your backyard is small or if you're concerned about noise and potential nuisances, bantam breeds might perfectly suit your needs. These smaller chickens require less space and tend to be quieter than their larger counterparts, reducing disturbances and allowing easy cohabitation with neighbors. For urban settings where noise can be a considerable issue, quieter breeds like the Australorp or even Silkies—though they require more grooming—can be a suitable choice, bringing charm without excessive noise.

Maintenance is another piece of the puzzle. If you're looking for breeds that thrive with minimal intervention, Sussex and Chantecler may be your best bet. These breeds are considered low-maintenance and adapt well to various environments without needing constant care, making them a favorable choice for beginners. On the other hand, designer breeds like Silkies require regular grooming due to their unique feathering. With their distinctive fluff and affectionate nature, they are often kept as pets rather than production birds but may not be ideal if you're looking for minimal upkeep.

To help you make an informed decision, I've included a breed comparison chart below. This chart visually lays out key traits like egg production, temperament, climate adaptability, space requirements, and maintenance needs. It's designed to guide you in selecting the perfect breed that aligns with your goals and environment.

BREED COMPARISON CHART

Breed	Egg Production	Temperament	Climate Adaptability	Space Requirement	Maintenance Needs
Rhode Island Red	High	Independent	Moderate	Standard	Low
Leghorn	Very High	Active	Moderate	Standard	Low
Buff Orpington	Moderate	Docile	High	Standard	Moderate
Australorp	High	Gentle	High	Standard	Low
Plymouth Rock	Moderate	Friendly	High	Standard	Low
Wyandotte	Moderate	Calm	High	Standard	Low
Bantams	Varies	Quiet	Moderate	Small	Low
Sussex	Moderate	Curious	Moderate	Standard	Low
Chantecler	Moderate	Hardy	High	Standard	Low
Silkies	Low	Friendly	Moderate	Small	High

This chart serves as a quick reference guide to help you compare and contrast potential breeds based on what's most important to you. Whether your focus is on egg production, ease of care, or simply having a group of chickens that get along well with each other and their human caretakers, choosing the right breed will set the stage for a successful chicken-keeping experience.

Imagine selecting your breeds as akin to building a small

symphony of characters in your yard, each contributing their melody to the daily rhythm of life. As the days go by, you'll find your chosen breeds reflecting aspects of your own lifestyle and community, blending into the fabric of your home life with ease.

WHERE TO ACQUIRE YOUR CHICKS

One of the first and most exciting steps is acquiring your first chicks. There are several reliable sources to consider, depending on your goals, budget, and the breeds you're interested in. Each source has its pros and cons, and knowing what to expect can help you make the best decision for your flock.

Local feed stores are a common starting point, especially in the spring. These stores often carry popular breeds of chicks that are suitable for backyard flocks. The benefit of buying from a local store is the ability to see the chicks in person, get advice from staff, and avoid shipping costs or stress on the birds. However, the selection may be limited, and it's important to verify that the chicks are vaccinated and properly cared for.

Another option is ordering chicks from a hatchery. Many hatcheries across the U.S. offer a wide range of breeds and will ship chicks directly to your door. This is a great way to get exactly what you want—whether you're looking for egg layers, meat birds, or rare heritage breeds. Hatcheries often have detailed descriptions and helpful guides, making the selection process easier. Just be aware that chicks are shipped through the postal service and must be picked up quickly to ensure their health and safety.

If you're looking for a more sustainable or community-based option, check local farms, homesteaders, or poultry swaps in your area. Sometimes these sources offer healthy, well-socialized chicks that are already acclimated to your local climate. Plus, connecting with experienced chicken keepers can provide invaluable knowledge and ongoing support as you get started. Whichever route you choose, be sure you're ready with the proper brooder setup and supplies before bringing your chicks home.

UNDERSTANDING LOCAL REGULATIONS AND ZONING LAWS

When I started my chicken-keeping research, I quickly discovered that the most significant hurdle wasn't building the coop or selecting the breeds—it was navigating the maze of local regulations and zoning laws. Just like any other pet, chickens come with their own set of rules and restrictions, often more complex than you might expect. To ensure you're on the right side of the law, it's crucial to research your local ordinances thoroughly. Begin by checking your city or county's official website, where you'll often find a section dedicated to animal and livestock regulations. These sites usually provide detailed information on what's allowed, including the maximum number of hens you can keep and whether roosters are permitted. Roosters are often banned in urban areas due to noise concerns, so if you're dreaming of a picturesque rooster crowing at dawn, you might need to adjust your plans.

In many regions, contacting your local agricultural extension office can be a goldmine of information. These offices are staffed by knowledgeable professionals who can provide insights into both state and local regulations. They often offer resources tailored for backyard poultry enthusiasts, ensuring that your flock complies with all health and safety standards. This step is especially important if you live in a densely populated area, where specific restrictions might apply to protect public health and minimize potential nuisances.

Permitting is another critical aspect to consider. Some areas require you to obtain a permit before you can bring chickens into your yard. The requirements for these permits can vary significantly, so be prepared to provide details about your planned setup, including coop dimensions and location. You may also need to demonstrate that you'll adhere to guidelines regarding waste management and cleanliness, essential for preventing unpleasant odors and potential health hazards.

If you're part of a homeowner association (HOA), you'll need to navigate an additional layer of regulations. HOAs often have their own rules about keeping animals, and chickens may not always be welcome. It's crucial to approach your HOA board with a well-prepared presentation highlighting the benefits of chicken-keeping. Focus on the positive aspects, such as sustainable living practices and the educational opportunities for children. Be prepared to address any concerns they might have about noise, odor, or potential property value impacts. Sometimes, demonstrating a commitment to maintaining a clean and discreet operation can make all the difference.

Once you've figured out all these regulations, maintaining legal compliance becomes your next priority. Keep copies of all relevant documentation organized in a dedicated folder or digital file—permits, correspondence with the city or HOA, and any other paperwork should be easily accessible in case you need to reference them quickly. Establishing good neighbor practices is also vital for staying in everyone's good graces. Let your neighbors know that you're planning to keep chickens and address any concerns upfront. Offering them fresh eggs from time to time can work wonders for fostering goodwill.

Remember that while keeping everything above board might seem daunting initially, it's an essential part of responsible chicken ownership. By diligently researching and understanding local regulations, not only do you protect yourself from fines and legal issues, but you also contribute positively to your community's perception of backyard chicken-keeping. Good legal standing allows you to enjoy your flock without worries, letting you focus on the rewarding aspects of this endeavor: fresh eggs on your breakfast table, the gentle clucking of contented hens, and the satisfaction of knowing you're living just a little bit closer to the earth.

The legal landscape is integral to the harmony within which you and your chickens coexist. By approaching this path with thoroughness and respect for regulations, you are paving the way for a sustainable, enriching relationship between your household and your feathered companions. Allow the process to guide you towards cultivating a neighborhood-friendly routine, one that emphasizes flourishing connections over hassles.

ESSENTIAL EQUIPMENT AND SUPPLIES FOR NEW CHICKEN OWNERS

Walking into the world of chicken-keeping, you'll quickly discover that outfitting your setup with the right gear is crucial. Just like any hobby, having the proper equipment makes all the difference. First and foremost, you'll need reliable chicken feeders and waterers. These are the lifelines for your flock, ensuring they have constant access to nourishment. Choose feeders that minimize spillage and waste, as chickens have a knack for scratching food everywhere. Gravity-fed waterers work well; they provide a steady flow of fresh water while preventing contamination. Investing in durable, easy-to-clean options will save you time and effort in the long run, allowing you to spend more moments enjoying your chickens rather than tending to upkeep.

Equally important are nesting boxes and roosting bars. Nesting boxes offer hens a cozy spot to lay their eggs, safeguarding them from damage and predators. Position these boxes slightly off the ground and fill them with soft bedding like straw or wood shavings to encourage egg-laying. Roosting bars, on the other hand, cater to a chicken's natural instinct to perch while sleeping. Elevate these bars within your coop to keep them safe from ground predators, ensuring a restful roosting experience for your flock. Consider the size of your chickens when installing these essentials; comfort is key to maintaining a happy, productive group of hens.

Now, let's talk about keeping things budget-friendly. If you love a good DIY project, creating your own feeder can be both economical and rewarding. A simple gravity feeder can be constructed using PVC pipes. This design not only saves money but also reduces feed waste by allowing chickens to peck at their leisure without spilling. You can also repurpose household items like old pie tins or shallow pans as temporary feeders or water dishes. Just ensure they are cleaned regularly to prevent contamination.

Safety cannot be overstated when it comes to keeping chickens. Predator-proof locks on the coop door are a must-have, especially if you're in an area with raccoon or fox activity. These cunning creatures can easily open simple latches, so opt for carabiner clips or padlocks to secure entries. Non-toxic cleaning supplies are another vital component of your safety gear arsenal. Chickens are sensitive to chemicals, so using natural cleaners like vinegar or baking soda not only keeps your coop sparkling but also protects your flock's health.

Creating a comprehensive checklist before acquiring your chickens ensures you don't overlook any details. This checklist should cover everything from basic gear to safety items:

CHICKEN-KEEPING ESSENTIALS CHECKLIST

- **Feeders**: Gravity-fed or DIY PVC feeders
- **Waterers**: Gravity waterers or automatic drinkers
- **Nesting Boxes**: One per 3-4 hens, filled with soft bedding
- **Roosting Bars**: Elevated perches for sleeping
- **Predator-Proof Locks**: Secure carabiner clips or padlocks
- **Bedding Material**: Straw, wood shavings, or sand
- **Non-Toxic Cleaning Supplies**: Vinegar, baking soda
- **Insulation Materials**: For cold climates
- **First Aid Kit**: Basic supplies for minor injuries

This checklist acts as a roadmap, guiding you through the initial setup phase and ensuring you're prepared for all aspects of chicken care.

Before you bring home your new flock members, take a moment to reflect on how you'll set up your space. Imagine this: It's a Sunday afternoon, and you're in your backyard setting up the final touches on your coop. You've got a sense of anticipation mixed with excitement because soon you'll be tending to your very own chickens. As you lay down fresh bedding and secure locks on the coop doors, you know these preparations will make all the difference in creating a safe haven for your chickens.

Remember that while you're focusing on practicality, there's room for creativity and personalization, too. Maybe you'll paint the coop in cheerful colors or add whimsical signs that reflect your style. These personal touches enhance not only the appearance but also your connection with this delightful endeavor.

The journey into chicken-keeping is as much about preparation as it is about discovery. Your newfound companions will provide endless opportunities for learning and joy. With the right equipment and a bit of creativity, you're well on your way to creating a haven where chickens thrive and eggs abound.

PLANNING AND DESIGNING YOUR CHICKEN COOP

As you step into the delightful realm of chicken-keeping, an important thing you'll want to consider is how to design a chicken coop that's both functional and inviting. We'll dive into this topic in greater detail in the next chapter, but for now, I want to give you some important points to consider.

Think of the coop as your chickens' home base—a sanctuary where they can eat, sleep, and lay eggs in safety and comfort. At its core, a well-designed coop needs to provide adequate space for each chicken. The general rule of thumb is about 2 to 4 square feet per bird inside the coop, with an additional 8 to 10 square feet in an outdoor run. This ensures they have enough room to move freely without feeling cramped. Adequate space prevents stress, which can lead to pecking and other behavioral issues.

Basic Chicken Coop

Ventilation is another critical element. Good airflow keeps the coop fresh and reduces the risk of respiratory diseases. You can achieve this by installing windows or vents near the roofline, allowing heat and moisture to escape without creating drafts at the birds' level. This setup is akin to the way we open windows in our homes to let in fresh air while maintaining a comfortable temperature. Ventilation not only keeps the air circulating but also helps control humidity levels, particularly in warmer climates, where heat can become oppressive.

When it comes to customizing your coop, the possibilities are endless. Perhaps you have a small yard and are worried about space. Portable coops might be your answer. These mobile units, often called "chicken tractors," allow you to move your chickens around the yard, giving them access to fresh grass while keeping them safe from predators. They're perfect for urban settings where space is at a premium. For those with larger flocks or more land, consider multi-tier designs that maximize vertical space. These designs not only provide ample room for roosting and nesting but also make egg collection more convenient.

Safety and security should remain at the forefront of your planning. Chickens are vulnerable to a host of predators, from raccoons and foxes to hawks and snakes. Reinforced wire mesh barriers are essential for keeping these threats at bay. Always choose hardware cloth instead of chicken wire, as the latter can be easily breached by determined intruders. Ensure all doors and openings have securely latching mechanisms—simple hooks or carabiners work well here —to prevent unauthorized entry by curious critters.

Incorporate user-friendly features to make daily tasks easier, such as external egg collection boxes that save you from entering the coop or feed storage compartments within easy reach. Little conveniences add up, making the routine of caring for your chickens less of a chore and more of an enjoyable part of your daily life. Having a well-thought-out coop design reflects not only efficiency but also the bond you are nurturing with your feathered friends.

To help inspire your coop design, consider practical layout examples that cater to different needs. Visualizing various setups can spark ideas on how best to arrange your coop based on your specific space and flock size. A simple rectangular layout with separate areas for nesting and roosting is a classic choice that suits most backyard flocks. Alternatively, L-shaped or circular designs can offer unique aesthetic appeal while still providing functionality.

Incorporating these elements into your coop design isn't just about meeting the basics; it's about creating an environment where your chickens can thrive. Picture this: It's a crisp morning, and as you sip your coffee, you watch your chickens cluck contentedly within their well-planned home. Their feathers glisten in the morning light as they peck around, exploring their surroundings with curiosity. The satisfaction of knowing you've provided a safe haven where they can express natural behaviors without fear is immensely rewarding.

As you embark on this planning phase, allow yourself the freedom to be creative yet practical. Your coop should reflect not only the needs of your chickens but also your personal style and the constraints of your environment. Personal touches like adding a small garden bed nearby or painting the coop in vibrant colors can transform a simple structure into a delightful backyard feature.

Remember that each decision you make now lays the groundwork for happy and healthy chickens. By ensuring ample space, effective ventilation, thoughtful customization, and robust security measures, you're setting the stage for success in your chicken-keeping endeavor. The journey of designing and building a coop may seem daunting at first, but with each step, you'll find yourself more connected to this fulfilling lifestyle choice—one that's not just about sustenance but about nurturing life and appreciating simplicity.

SETTING UP A SAFE AND COMFORTABLE BROODER

Setting up a brooder for your chicks is like crafting their first cozy home. It's where they'll eat, sleep, and grow until they're ready to join the others in the coop. The brooder is essentially a warm, secure space that mimics a mother hen's care. Start with the heat source, which is vital for young chicks who can't regulate their body temperature. Heat lamps are a popular choice due to their efficiency and reliability. Hang the lamp securely above the brooder, ensuring it's safely out of reach from flammable materials. Alternatively, consider using a radiant heat plate—a safer option that replicates the warmth of a mother hen by allowing chicks to huddle beneath it. Both methods will keep your chicks snug and comfortable.

Cutaway Image of a Brooder

Now, let's talk bedding. Bedding absorbs waste, reduces odor, and offers a soft surface for the chicks to walk on. Pine shavings are an excellent choice—affordable, absorbent, and easy to clean. Avoid cedar shavings; they can be toxic to chicks. Some people use straw or shredded paper, but these can be less absorbent and might require more frequent changes. Layer the bedding about two inches thick over the floor of the brooder to ensure adequate cushioning.

Maintaining the right temperature and humidity in the brooder is crucial for chick health. During their first week, chicks need a cozy environment of about 95°F. Reduce this temperature by 5°F each week until they're ready for the coop. A simple thermometer placed at chick level will help monitor this precisely. Pay attention to humidity, too—keeping it around 50% is ideal. Use a hygrometer to track humidity levels, adding a shallow dish of water to increase moisture if needed.

Brooder maintenance involves regular cleaning to prevent disease and promote healthy growth. Chicks are messy little creatures, so establish a cleaning schedule early on. Remove droppings daily and change bedding at least once a week. Keeping surfaces clean is essential for preventing respiratory issues. Disinfect the brooder every few weeks using a solution of water and vinegar or mild soap—avoid harsh chemicals that could harm your chicks.

Monitoring chick behavior gives you invaluable insight into their well-being. Chicks naturally chirp and explore, but excessive peeping might indicate discomfort. If you notice them huddling under the heat source, they're likely cold. Conversely, if they're scattered far from it, they might be too warm. Adjust the heat source accordingly to maintain their comfort zone. Regularly observe their activity and feed consumption; these are indicators of health and contentment.

Keeping an eye out for signs of illness is also crucial. Healthy chicks are active with bright eyes and clean, fluffy feathers. Look for symptoms such as lethargy, labored breathing, or drooping wings—these may suggest health issues requiring prompt attention. If you spot any signs of illness, separate the affected chick immediately to prevent spreading disease and consult with a vet experienced in poultry care.

Establishing a nurturing environment within the brooder is essential for laying the groundwork to raise strong, healthy chickens. As each day passes, you'll observe not just physical growth but also the development of a deeper bond between you and your chicks. This formative stage is critical, as it's during this time that these tiny, delicate creatures gradually evolve into hearty, robust birds, fully equipped and prepared for the broader world beyond the confines of their first home. Watching this transformation unfold is a remarkable journey, filled with moments of learning, care, and connection that underscore the importance of your role in their early life. This nurturing process ensures that by the time they are ready to transition to the coop, they are not only physically prepared but also have developed a trust in their environment and in you, their caretaker, setting the stage for a successful integration into the flock.

In crafting your brooder setup, remember that flexibility is key. Each flock has its quirks and preferences, so adjusting based on their behavior will ensure you meet their needs effectively. It's all about finding that sweet spot where science meets instinct—creating a space that's both practical and comforting for your chicks as they take their first steps in life.

INTRODUCING YOUR FIRST FLOCK TO THE BACKYARD

Bringing your chickens into their new outdoor environment is a moment of excitement and anticipation. It's akin to watching children explore a playground for the first time. But just like kids, chickens need time to acclimate to their surroundings gradually. When you first introduce them to the backyard, short, supervised visits are crucial. Start by allowing them to explore a small, contained area where they can get used to the sights, sounds, and smells of their new world. These initial outings should be brief, perhaps just ten to fifteen minutes, slowly increasing as they become more comfortable and confident.

As your chickens grow accustomed to their new environment, you can gradually expand their roaming area. This gradual expansion helps prevent overwhelming them and allows them to establish a sense of security. It's akin to giving them training wheels before they're ready to ride freely. You'll notice that, over time, they'll begin to venture further and further, pecking and scratching with increasing curiosity. This process not only aids in their acclimation but also helps them develop a healthy routine of foraging and exploring.

Flock dynamics play a significant role in how your chickens interact with one another. Chickens naturally establish a pecking order—a social hierarchy that determines dominance within the group. It's fascinating to observe these dynamics unfold as they assert their roles through subtle behaviors. You might see one chicken strutting confidently while others defer by stepping aside. This is normal behavior and usually resolves itself without intervention. However, if aggression becomes an issue, there are strategies to minimize it, and we will address this topic thoroughly later in the book.

Predator awareness is an essential aspect of keeping your flock safe. Every backyard has its own set of potential threats, ranging from neighborhood cats and dogs to more elusive foes like raccoons and hawks. Recognizing these local predators is the first step in protecting your flock. Pay attention to what's prevalent in your area —ask neighbors or check local wildlife resources for common threats. Implement deterrent measures, such as installing motion-activated lights or decoy owls to discourage predators from approaching. Secure fencing and covered runs provide additional layers of protection.

Sharing stories from fellow chicken keepers can offer both motivation and reassurance. Take, for instance, the experience of a friend who initially struggled with integrating a new flock into her busy urban backyard. She started with short outings, just as we've discussed, but noticed her flock was hesitant to explore beyond their coop. With patience and a few enticing treats scattered around their enclosed area, she gradually encouraged them to roam further each day. Within a couple of weeks, her chickens were confidently strutting across the yard, engaging in natural behaviors like dust bathing and foraging.

Another fellow chicken keeper faced challenges with predator threats in his rural setting. He discovered raccoons frequenting his property at night. By using solar-powered predator lights that mimic eyes glowing in the dark, he effectively deterred these nocturnal intruders without harming them or his chickens. These real-life accounts illustrate that while challenges may arise, there are always creative solutions to navigate them successfully.

Remember that each chicken keeper's experience is unique, shaped by individual circumstances and environments. But the underlying principles of gradual acclimation, understanding flock dynamics, and ensuring predator protection are universal keys to success. As you embark on this adventure with your flock, embrace the learning curve with openness and curiosity. You'll find that watching your chickens thrive in their new home brings a sense of accomplishment and joy that few other hobbies can match.

As you conclude this foundational chapter in your chicken-keeping adventure, reflect on the steps you've taken so far and the knowledge you've gained. The initial stages of introducing your flock are filled with discovery—not just about chickens but also about yourself as a caretaker. Each moment spent observing their antics and nurturing their growth reinforces the bond between you and these feathered companions. It's a rewarding experience that promises fresh eggs along with endless moments of amusement. With patience, care, and an open heart, you are well on your way to becoming a confident and capable chicken keeper.

CHAPTER 2
BUILDING THE PERFECT COOP

O n a breezy afternoon, I found myself researching coop designs, my mind brimming with ideas. It was then I realized that creating a safe haven for my flock was more than just a construction project—it was a labor of love. You see, the coop isn't merely a shelter; it's a fortress protecting your chickens from the sly antics of neighborhood predators. Crafting a predator-proof coop demands attention to detail and an understanding of potential threats.

UNDERSTANDING THE BASICS

A well-designed chicken coop is essential for keeping your flock safe, healthy, and productive. At the heart of the setup is the coop itself, a secure, weatherproof shelter where chickens can sleep, lay eggs, and take refuge from predators and harsh weather. The coop should be well-ventilated but draft-free, and it must provide enough space for each bird—generally about 2–4 square feet per chicken inside the coop. A solid floor, tight-fitting doors, and hardware cloth over any openings help keep out predators like raccoons, snakes, and weasels.

Connected to the coop is the chicken run, an outdoor area where chickens can roam, scratch, peck, and enjoy the sunlight. This space should be enclosed with strong wire or mesh to prevent escapes and deter predators, particularly from above.

Inside the coop, nesting boxes are essential for egg-laying hens; these should be filled with clean bedding and placed in a quiet, dim area of the coop to encourage hens to lay there consistently. Each box can serve about 3–4 hens.

Finally, roosting bars are critical for nighttime, as chickens prefer to sleep off the ground. These bars should be placed higher than the nesting boxes to prevent chickens from sleeping where they lay eggs, and they should be wide enough for the birds to comfortably grip while they sleep. Together, these components create a safe and functional environment for your backyard flock.

Interior of a Coop, with Nesting Boxes and Roosting Bars

BUILDING OPTIONS

Many readers will be highly experienced in various types of DIY projects, and may be eager to take on the challenge of designing and building a chicken coop from scratch. I will share tips for the DIYers later in this chapter. Others might find the prospect of doing this on their own from the ground up more daunting. Don't worry, I've got you covered.

Here are two excellent online resources that provide detailed plans and instructions for building a chicken coop, ranging from beginner-friendly to more advanced builds:

1. Backyard Chickens – Coop Plans & Designs

- **URL:** https://www.backyardchickens.com/articles/category/chicken-coops.12/
- **What it offers:** Hundreds of user-submitted chicken coop plans with photos, materials lists, and step-by-step instructions. Great for all skill levels.
- **Why it's useful:** Real-life examples from chicken keepers, with feedback and variations.

2. The Garden Coop

- **URL:** https://www.thegardencoop.com
- **What it offers:** Professionally designed chicken coop and run plans available for purchase, including the popular "Garden Coop" and "Garden Ark."
- **Why it's useful:** Well-designed plans with precise instructions, diagrams, and tips—ideal for secure, predator-proof coops.

And if you tend to shy away from DIY building projects altogether, fear not, there are sources for prefabricated options as well. Here are two of the best.

1. Omlet

- **Website:** https://www.omlet.us
- **Direct link:** https://www.omlet.us/chicken-coops/
- **What they offer:** Modern, weatherproof, and predator-resistant chicken coops like the *Eglu Cube* and *Eglu Go*. Modular designs with options for wheels, runs, and accessories.
- **Why it's great:** Known for durability, easy cleaning, and minimal maintenance. Great for urban or suburban backyards.

2. Tractor Supply Co.

- **Website:** https://www.tractorsupply.com
- **Direct link:** https://www.tractorsupply.com/tsc/catalog/coops-pens
- **What they offer:** A wide selection of prefab wooden chicken coops, ranging from small backyard setups to larger walk-in coops.
- **Why it's great:** Offers in-store pickup or delivery. Good variety in budget, style, and size.

STEPS TO BUILDING A PREDATOR-PROOF COOP

First and foremost, let's delve into the materials you'll need to make your coop as secure as possible. Hardware cloth is your best friend here. Unlike chicken wire, which can be easily breached by determined predators, hardware cloth offers robust protection. Its small mesh size prevents sneaky paws from reaching in and causing harm. Make sure to cover all windows and vents with hardware cloth, securing it tightly to avoid any gaps that crafty critters might exploit. As an added layer of security, consider using heavy-duty staples or screws with washers rather than nails, which can be pried loose over time.

Next on the list, consider installing underground barriers to deter digging animals, like raccoons and foxes. Burying hardware cloth or concrete blocks about a foot deep around the perimeter of your coop creates an impenetrable barrier. This extra step might seem tedious, but it's worth every bit of effort when you see your chickens safe and sound. To further enhance this, some chicken keepers have found success by including sharp-edged materials or coarse gravel beneath the barrier, adding a dissuasive texture that makes digging uncomfortable for potential predators.

When it comes to construction materials, pressure-treated wood is a solid choice for durability. It withstands the elements and stands firm against the test of time. However, it's crucial to avoid using wood treated with chemicals harmful to chickens if they decide to peck it. Non-toxic finishes are available and should be used to ensure their safety. Ensure all doors are equipped with reinforced locks and latches. Opt for two-step latches or carabiners that raccoons can't figure out—because they will try! A determined predator can open simple locks, so this added security measure is crucial.

Now, let's delve into some design strategies that naturally deter unwanted visitors. Elevating your coop is an effective way to keep ground predators at bay. By raising the structure off the ground, you eliminate easy access points for digging animals, providing an additional layer of security. Moreover, incorporating a slanted roof not only aids in water runoff but also helps prevent climbing creatures from gaining a foothold. This simple architectural feature can significantly reduce the risk of predators scaling your coop. For a more efficient design, solar or motion-activated deterrent lights can be installed at vantage points to frighten nocturnal invaders.

PRACTICAL EXAMPLES AND REAL-WORLD TECHNIQUES

Real-world experiences often provide invaluable lessons. I recall a fellow chicken keeper who initially underestimated the persistence of local raccoons. After several frustrating attempts to secure her coop with basic locks and chicken wire, she switched to hardware cloth and installed underground barriers. The transformation was remarkable—no more missing chickens or midnight disturbances. Her story is a testament to the effectiveness of these preventive measures. To enhance her setup, she later added perimeter alarms that alert her when disturbances occur, offering real-time protection.

Whether or not you live in a predator-heavy area, investing time in these protective strategies ensures long-term peace of mind. A friend I know, who also faced numerous predator challenges, added additional layers of security by installing motion-activated floodlights around the coop's perimeter. These lights, coupled with his existing reinforcements, became a powerful deterrent. Knowing that predators would flee the startling sudden brightness provided him peace and his chickens tranquility. As dusk settles, these lights activate in gentle increments to avoid startling the chickens but effectively ward off unwanted visitors.

CASE STUDY: LESSONS LEARNED FROM PREDATOR-PROOFING SUCCESSES

Imagine your coop as Fort Knox for chickens. Another friend of mine, living in a rural area with frequent predator sightings, faced numerous challenges before achieving success. Initially, he also used chicken wire and basic latches, only to discover raccoons easily breached them. After switching to hardware cloth and carabiner clips, his flock remained undisturbed. The small investment in these security measures paid off in peace of mind and happy chickens. Furthermore, he introduced metal flashing around the base of the coop, preventing rats from gnawing their way inside. It was a revelation, showing just how intricate a predator's attempts can be and how prepared one must be to outsmart them. His journey underscored the principle that in predator-proofing, redundancy isn't just a precaution—it's a necessity.

An additional consideration he implemented was a perimeter noise deterrent system that emitted high-pitched frequencies only audible to animals. This innovative approach drastically reduced the number of predator visits, adding another successful layer to his protection strategy.

By following these guidelines and learning from real-life examples, you can build a coop that not only meets your needs but also stands strong against potential threats. Your chickens will show their appreciation with clucks of contentment as they thrive in their safe haven. Remember that each step you take in fortifying your coop serves as an investment in their well-being, allowing you to enjoy the many rewards of backyard chicken-keeping without worry.

OPTIMAL COOP VENTILATION AND TEMPERATURE CONTROL

Imagine this: it's a hot summer day, and your coop feels like an oven. The air is heavy, and your chickens are panting, desperately seeking relief. Proper ventilation is crucial in preventing such scenarios. It ensures your flock remains healthy by allowing fresh air to circulate, expelling moisture, and keeping harmful bacteria at bay. Place windows or vents high up to let warm air escape without creating drafts at the birds' level. In warmer regions, consider installing fans or wind-driven turbines to enhance airflow. These simple additions can make a world of difference in keeping your coop fresh and your chickens comfortable. To monitor the efficiency of your ventilation system, integrate humidity and temperature sensors that offer real-time data to adjust ventilation accordingly.

Temperature control is another key factor in maintaining a healthy coop environment. Insulated walls can be a game-changer, especially in regions with extreme weather conditions. They keep the coop cool in summer and warm in winter. For those chilly months, a heat lamp can offer much-needed warmth, but safety is paramount. Ensure it's securely placed, away from flammable materials, to prevent accidents. Consider using a ceramic heat emitter instead of traditional bulbs. They provide consistent heat without the risk of shattering. Some keepers vouch for heated flooring options, which evenly distribute warmth and alleviate the need for overhead heat sources, providing another point of comfort for your chickens.

STRATEGIES TO MANAGE MOISTURE AND AIR QUALITY

Managing moisture levels is essential to prevent respiratory issues in chickens. High humidity can lead to damp bedding and mold growth, creating an unhealthy environment. Drip edges and gutters on the roof can effectively direct rainwater away from the coop, minimizing moisture accumulation. Inside, choose bedding materials like straw or wood shavings that absorb moisture well and can be changed easily. Regularly check for wet spots and replace damp bedding promptly to maintain a dry and comfortable living area for your flock. Implement moisture-absorbing pouches or dehumidifiers in especially rainy climates to combat high humidity effectively.

Insights from seasoned chicken keepers and veterinarians underscore the importance of environmental control. One experienced keeper shared how she battled respiratory issues in her flock by tweaking ventilation and moisture management strategies. By adding extra vents and switching to sand bedding, she noticed a significant improvement in her chickens' health. Veterinarians also emphasize regular air quality checks, recommending that coop owners pay attention to ammonia levels—a common byproduct of chicken droppings that can irritate respiratory systems if allowed to build up. Installing ammonia monitors regularly within the coop can alert you to unhealthy levels, allowing for prompt action to ensure the well-being of your flock.

Incorporating these expert insights into your coop setup can significantly impact your chickens' well-being. It's not just about following guidelines; it's about observing your flock and making necessary adjustments to ensure they thrive. For instance, if you notice condensation on the windows or a musty smell, these are signs of inadequate ventilation or high humidity levels that need immediate attention.

A well-ventilated coop with stable temperature control will not only enhance your chickens' comfort but also boost egg production and overall flock health. The effort you put into creating an optimal environment reaps rewards in the form of happy, productive birds. As you fine-tune your coop's setup, remember that small changes can lead to significant improvements. Each adjustment you make brings you closer to providing the best possible home for your feathered friends.

Your chickens rely on you to create a space where they can thrive, lay eggs consistently, and live comfortably. By focusing on ventilation, temperature regulation, and moisture control, you're setting the stage for success in your chicken-keeping endeavors. Ultimately, it's about creating a balanced environment that caters to your flock's needs while ensuring their safety and happiness.

CREATING A SECURE OUTDOOR RUN FOR YOUR CHICKENS

Imagine your chickens enjoying the fresh air and sunshine in their secure outdoor run, happily pecking and scratching away. It's crucial to provide them with a safe space to roam, where they can indulge in their natural behaviors while remaining protected from potential threats. When constructing your chicken run, one of the first decisions you'll face is choosing the right fencing material. Welded wire offers a sturdy barrier that keeps most ground predators at bay. Its rigid structure and small mesh size prevent unwanted guests from squeezing through gaps, ensuring your flock's safety. On the other hand, electric fencing provides an extra layer of security, delivering a mild shock to deter curious critters. It's an effective option for those living in areas with persistent predator problems.

DESIGNING ACCESS POINTS AND ENHANCING SECURITY

Designing access points is another key consideration when setting up your run. You'll want to create entrances that are easy for you to use but challenging for predators to breach. Consider installing a simple latch system that allows you to open the gate with one hand while carrying feed or water with the other. Make sure the gates are secure and fit snugly within the frame to prevent any sly creatures from slipping through. A double-door entryway can add an extra level of security, giving you peace of mind knowing your chickens are safe. Implement technology such as automated locks that respond to your presence or specific times, adding a high-tech touch to your chicken care practices.

While solid fencing forms the backbone of your run's defenses, additional deterrents can further increase its security. Motion-activated lights or sprinklers serve as excellent surprise tactics against nocturnal predators. A sudden flash of light or burst of water can startle and discourage them from approaching your chickens' sanctuary. Overhead netting is crucial for protecting against aerial predators like hawks and owls. By covering the top of your run with durable netting, you prevent these birds of prey from swooping down and making off with one of your hens.

Beyond safety, it's important to enrich your chickens' environment to stimulate their natural behaviors. Adding areas for dust bathing is a must; chickens love nothing more than rolling around in loose sand or peat, cleaning their feathers and keeping parasites at bay. These little dust baths can be as simple as a shallow container filled with sand or an area of the run dedicated to this purpose. Providing perches and climbing structures encourages exercise and satisfies their instinctual need to roost above ground level. Repurposing branches or constructing wooden ladders gives them something to climb on, promoting physical activity and mental stimulation.

To help visualize your setup, it might be helpful to consider different layout options for efficiency and security. Picture a rectangular run with a central access path, allowing easy reach to all areas for cleaning and maintenance. The perimeter could feature reinforced fencing paired with strategically placed deterrents like lights or sprinklers at each corner. Inside this safe zone, designated areas for dust bathing alongside scattered perches create an enriching environment that caters to your chickens' needs. Enhance the landscape with varying levels, such as gentle inclines or raised platforms, providing your chickens with stimulating topography similar to their natural habitat.

As you plan your outdoor run, remember that each choice you make contributes to both the safety and happiness of your flock. No detail is too small when it comes to ensuring their well-being; even the material you choose for the ground can have significant impacts. Consider a mix of gravel and sand for the flooring of your run. Gravel promotes drainage, which helps keep the area dry after rain, while the sand provides a comfortable substrate for dust bathing. Balancing these elements ensures not only their well-being but also enhances your experience as a chicken keeper. Watching your chickens thrive in a secure environment brings immense satisfaction and joy, knowing you've created a space where they can live freely yet safely.

Establishing a secure and enriching outdoor run allows your chickens to explore their surroundings confidently while remaining protected from potential hazards. This harmonious blend of security measures and environmental enrichment fosters a thriving backyard flock that rewards you with contentment and productivity.

DIY COOP DESIGNS AND CREATIVE SOLUTIONS

Building a chicken coop offers a wonderful opportunity to let your creativity shine. It's a chance to tailor a space that meets both the needs of your chickens and your personal style. One way to make your coop stand out is by using recycled materials. Not only does this approach save money, but it also promotes sustainability. Old wooden pallets, for instance, can be deconstructed and repurposed into sturdy walls or flooring. They provide a rustic charm while being budget-friendly. With a little imagination, you can transform these simple materials into a charming coop that blends seamlessly with your backyard.

PERSONALIZATION AND COST-EFFECTIVE DESIGN

Personalization comes into play when you add decorative touches like paint or murals. Imagine bright colors or whimsical designs adorning your coop, turning it into a backyard centerpiece. A splash of color can bring joy to your outdoor space, and painting the coop with themes or patterns makes it uniquely yours. Some folks even involve their kids in the painting process, creating a fun family project that everyone can enjoy. It's amazing how a few brushstrokes can breathe life into a simple structure, making it not just a coop but a piece of art.

For those on a tight budget, inventive solutions abound. Pallet wood isn't just for walls; it can be used to construct the entire framework of your coop. This method is cost-effective and still allows for plenty of customization. Repurposing old furniture as nesting boxes is another clever idea. An unused dresser, for example, can be converted into cozy nesting spaces by removing the drawers and adding bedding. These creative hacks ensure functionality without breaking the bank, demonstrating that resourcefulness is key in building an affordable yet efficient coop.

Space constraints? No problem! Modular coops offer an ideal solution. These designs allow you to expand as needed, adding sections or levels as your flock grows. It's like building blocks for adults, giving you the flexibility to adapt your coop over time. Integrating your coop with existing structures, like garden sheds, is another smart move for maximizing space. By sharing walls or roofs, you reduce material costs and create a seamless connection between different areas of your yard.

Let's not forget the brilliant ideas submitted by fellow chicken keepers who have mastered the DIY approach. One reader shared their success with a vertical coop design that utilizes minimal ground space while providing plenty of room for chickens to roost and nest above eye level. This design is perfect for urban settings where land is limited. Another reader crafted a charming coop using reclaimed barn wood, creating a rustic yet functional space that blends beautifully with its surroundings. These examples show that with a little ingenuity and effort, you can create something truly special.

Incorporating natural elements into your design can also enhance the coop's aesthetic appeal and functionality. Use branches or logs for perches inside the coop, adding texture and variety to the environment. These natural materials are not only cost-effective but also encourage chickens to exercise their natural behaviors by climbing and perching at different heights. Introduce small plants or herbs around the coop's exterior; certain herbs, like lavender or mint, help deter pests and add a delightful fragrance to the air. Creating a DIY chicken coop is about more than just providing shelter; it's about creating a space that reflects your personality and meets the needs of your chickens. Whether you're repurposing materials or adding artistic touches, each decision contributes to a unique and inviting environment. The process of building your coop becomes an adventure in itself, filled with opportunities for innovation and personal expression.

As you embark on this creative endeavor, remember that each coop is as unique as its owner. Your choices in materials, design, and decoration breathe life into the structure, transforming it from mere shelter into a cherished part of your home landscape. With thoughtful planning and a touch of imagination, your DIY coop will stand as a testament to both your ingenuity and your commitment to sustainable living.

Your chickens will appreciate the effort you've put into creating their home, responding with contented clucks and fresh eggs as thanks for providing them with such a delightful abode. This balance of practicality and creativity ensures that both you and your flock thrive, enjoying the fruits of your labor in a space that's entirely your own.

ESSENTIAL COOP MAINTENANCE TIPS FOR BEGINNERS

Waking up to the soft clucks of your chickens and the promise of fresh eggs is a wonderful way to start the day. But to keep your coop running smoothly, a little maintenance goes a long way. Let's break down a simple routine that keeps your coop clean and healthy without overwhelming you. On a daily basis, remove droppings from under the roosting bars. This quick task helps reduce odor and keeps pests at bay. Freshen up the water supply, and make sure your chickens have plenty of clean water to drink. A simple rinse of the waterer can prevent any buildup of bacteria, which is crucial for maintaining flock health. These small daily habits ensure your flock stays happy and your coop remains pleasant.

Weekly, take some time to inspect your coop for any signs of damage. Look for loose boards, gaps in walls, or anything that might compromise its security. Pests can find tiny openings to sneak through, so sealing these up immediately prevents bigger problems down the road. Weekly checks also involve looking for pest infestations. Mites and lice are tiny but can cause big problems if left unchecked. Observe your chickens for signs of over-preening or feather loss, as these might indicate an unwelcome invasion. A proactive approach keeps your chickens comfortable and reduces stress, leading to better egg production.

When it comes to effective cleaning methods, natural options like vinegar or baking soda are your best bet. Vinegar is excellent for disinfecting surfaces without exposing your chickens to harsh chemicals. Mix it with water for a simple spray that can be used on walls and nesting boxes. For deeper cleaning sessions, pressure washing can be immensely effective. It blasts away grime and stubborn dirt from the coop's exterior, leaving it looking fresh and new. Just be sure to remove your chickens first and allow the area to dry completely before letting them back in. This method saves time and effort, making thorough cleanings less of a chore.

PREVENTIVE AND SEASONAL MAINTENANCE

Keeping your coop in tip-top shape involves knowing how to spot signs of wear and tear before they become major issues. Regularly check hinges, latches, and locks to ensure they work smoothly. Seasonal maintenance is equally important, particularly as the weather changes. In the fall, prepare for winter by checking insulation and sealing any drafts that could let in cold air. Weather strips can be an inexpensive and highly effective way to further seal these gaps. In spring, reinforce areas that might have weakened during winter storms or heavy rains. Paying attention to these details fortifies your coop against the elements, preserving its structure and protecting your flock.

Troubleshooting common maintenance issues doesn't have to be daunting. Leaks are often caused by aging roof materials or clogged gutters. If you spot water inside the coop after a rainstorm, inspect the roofline for gaps or cracks and patch them up promptly. Drafts can be tricky; they often sneak in through unnoticed crevices. Feel around on a windy day to locate drafts and seal them with weatherproof tape or caulking. Handling infestations requires diligence and quick action. If mites or lice are discovered, a thorough cleaning followed by dusting your chickens with diatomaceous earth can help eradicate these pests naturally.

A checklist can be a lifesaver when it comes to keeping track of maintenance tasks. Consider drafting one that outlines daily, weekly, and monthly responsibilities, so nothing falls through the cracks. This checklist guides you through routine upkeep while acting as a reminder for less frequent tasks like deep cleaning or structural inspections.

MAINTENANCE CHECKLIST

- **Daily:** Remove droppings, refresh water supply
- **Weekly:** Inspect for damage, check for pests
- **Monthly:** Deep clean with vinegar solution or pressure wash
- **Seasonal:** Weather-proofing adjustments (seal drafts in fall, reinforce in spring)

Embracing routine maintenance ensures your coop remains a safe haven for your chickens. It might seem like a lot at first glance, but once these tasks become part of your regular schedule, you'll find they blend seamlessly into your day-to-day activities. The satisfaction of knowing you've created and maintained a clean, secure environment for your flock is incredibly rewarding. Whether you're scrubbing down walls or tightening loose screws, each effort contributes directly to the well-being of your chickens—and that makes every bit of work worthwhile.

INCORPORATING NATURAL LIGHTING AND SHADE

Over time, I've come to realize the immense benefits of natural lighting for our feathered friends. Natural light plays a pivotal role in stimulating egg production and promoting overall chicken health. Positioning windows strategically in your coop can harness this light effectively. If you live in the Northern Hemisphere like me, aim to install windows on the south-facing side, where they capture the most sunlight throughout the day. This positioning ensures your chickens enjoy the benefits of natural light without overheating.

IMPLEMENTING LIGHT AND SHADE SOLUTIONS

Installing skylights or clear roof panels can further enhance this effect, bathing your coop in gentle, diffused sunlight that mimics their natural environment. Chickens are incredibly perceptive to light, and these features can help regulate their laying cycles and mood. It's remarkable how a few well-placed windows or panels can transform the coop into a sunlit haven, encouraging your hens to lay consistently and remain active and healthy.

However, too much sun exposure can lead to heat stress, especially during those blazing summer months. Providing ample shade is crucial for keeping your flock comfortable. Planting trees or shrubs around the coop offers natural shade that cools the area while adding beauty to your backyard. The foliage not only provides shelter from the sun but also creates a natural habitat for insects, adding a touch of biodiversity that your chickens will love exploring.

When immediate shade is needed, consider using shade cloths or tarps for temporary coverage. These versatile solutions can be draped over the coop and run, offering respite from the sun's intensity. They are easy to adjust and can be removed when the weather cools down. Combining permanent and temporary shading solutions ensures your chickens have access to cooler areas throughout the day, reducing their stress levels and promoting wellbeing.

In the evenings or during darker days, energy-efficient lighting alternatives come into play. Solar-powered lights offer an eco-friendly option for illuminating coops without additional electricity costs. Placing them strategically around your coop can provide just enough light for evening checks or late-night egg collection. LED bulbs are another sustainable choice, offering bright illumination with minimal energy consumption. These modern options help maintain a gentle balance between natural and artificial lighting, ensuring your flock enjoys a consistent environment. Automated timers can regulate artificial lighting, simulating the natural sunrise and sunset to minimize disruption to your flock's circadian rhythm.

Taking inspiration from fellow chicken keepers can provide valuable insights into effective lighting and shading solutions. A neighbor of mine cleverly utilized old window frames fitted with clear polycarbonate sheets as skylights in her coop. The result was a bright, airy space that stayed cool, even during peak summer heat. Another friend planted fast-growing shrubs around her run, creating dappled shade that transformed her chicken yard into a lush oasis. These real-world examples demonstrate that with creativity and a bit of effort, you can achieve an ideal balance of light and shade in your setup.

By considering both natural light and shading techniques, you create an environment where your chickens thrive. They enjoy the benefits of sunlight for health and egg production while having cool retreats to escape the heat. Balancing these elements is key to fostering a happy and productive flock.

As we wrap up this chapter on building the perfect coop, remember that every decision you make impacts your chickens' quality of life. From predator-proofing to optimizing lighting and shade, each aspect contributes to creating a safe, comfortable home for your flock. With these foundational elements in place, you're ready to explore the next chapter: feeding and nutrition essentials. This new knowledge will further enhance your ability to care for and enjoy these delightful creatures.

CHAPTER 3
FEEDING AND NUTRITION ESSENTIALS

UNDERSTANDING CHICKEN DIETARY NEEDS AND OPTIONS

Witnessing the shimmering plumage and sprightly demeanor of my neighbor's chickens serves as an everyday testament to how profoundly a balanced diet influences poultry vitality. Just as diverse, nutrient-rich meals cater to human health, chickens thrive on diets loaded with essential nutrients that fuel their growth and overall well-being. A well-constructed diet can specifically increase their lifespan, improve their resilience against diseases, and enhance their productivity, reflecting how indispensable good nutrition is to their maintenance and management.

At the heart of a healthy chicken diet is protein, a critical building block facilitating robust muscle and feather development. Protein-rich feeds ensure not just physical growth but also energy for daily activities. Chickens engage in constant movement throughout the day, pecking at the ground and flapping their wings. This activity necessitates a steady supply of protein to replenish and repair muscle tissue, thereby acting as the bedrock of their energetic lifestyle. Meanwhile, calcium is indispensable for robust eggs, fortifying eggshells to prevent the disappointment of fragile, thin-shelled eggs that compromise egg quality and safety. Without adequate calcium, hens may suffer from weak bones, as the body will redirect calcium reserves to shell formation, jeopardizing their skeletal structure in the process and thereby compromising long-term health.

Furthermore, vitamins A, D, and E serve critical roles akin to guardians who bolster the immune system, ensuring that daily threats from pathogens remain at bay. Vitamin A is particularly crucial for maintaining healthy eyesight and skin, influencing the iridescent sheen of their feathers, which is often an indicator of a well-maintained chicken. Vitamin D, often synthesized through sunlight exposure, enhances the uptake of calcium, thereby serving a dual purpose in maintaining egg integrity and skeletal health. Vitamin E acts as an antioxidant, neutralizing free radicals, thus helping mitigate oxidative stress in chickens' bodies, which is vital to preserving their overall vitality and longevity.

As juvenile chicks journey from hatching to full-fledged hens or roosters, their nutritional needs undergo significant transformations. Chicks initially thrive on starter feed, which is abundantly rich in protein to accommodate their rapid initial growth and feather development. Starter feeds are often formulated to be easily consumable, with the grains and protein pellets finely ground to ensure chicks, whose pecking abilities are still developing, can consume meals efficiently. This dietary stage lays the groundwork for enduring health, imprinting well-being that echoes throughout their lives. As maturity beckons, their diet transitions towards layer feeds predominated by higher calcium levels. This adaptation coincides with their burgeoning roles as egg layers, heralding their contribution to egg production, which could potentially grace family breakfasts with their offerings.

Exploring feed options reveals a spectrum of possibilities, each catering to different farming philosophies and objectives. Organic feeds appeal to those championing pesticide-free, non-GMO nutrition. Such feeds are often perceived to enhance the overall quality of eggs and meat due to the absence of synthetic chemicals, potentially reducing the health risks associated with these additives. Organic feed options include ingredients like whole grains, fish meal, kelp, and unrefined barley malt, integrating seamlessly into an organic course of living. However, these luxurious options often come with a heftier price tag and may not be as accessible throughout all regions, thereby posing a dilemma for budget-conscious poultry keepers.

In contrast, commercial feeds offer a more approachable solution, usually fortified with a comprehensive array of essential minerals that promise balanced nutrition. Their affordability and broad availability make them particularly attractive, though some contain additives that certain keepers might prefer to eschew. It is imperative for poultry raisers to carefully scrutinize ingredient labels to avoid unwanted components, particularly for those with commitments to ethical farming or natural diets.

For enthusiasts eager to delve into the intricacies of advanced feeding methods, fermented feed arises as an intriguing alternative. Grains soaked until fermentation optimize digestibility and enhance nutrient bioavailability, paving the way for improved assimilation. Fermented feeds cultivate beneficial gut flora, leading to better digestion and improved egg production outcomes. Though this method demands patience and prior planning, involving placing grains in water and allowing them to sit for several days while stirring occasionally, it rewards caretakers with observable enhancements in the health and vitality of the flock. Fermentation not only enhances the nutritional profile but also reduces phytic acid levels present in grains, which otherwise may hinder mineral absorption.

NUTRITIONAL CHECKLIST

- **Protein Sources**: Fish meal, soybean meal, mealworms, peanuts
- **Calcium Supplements**: Crushed oyster shells or limestone
- **Vitamins**: A (carrots), D (sunlight exposure), E (spinach, nuts, and seeds)
- **Fiber Options**: Alfalfa hay, chia seeds, sunflower seeds

Navigating through the vast array of nutritional choices is both an exploration and a learning journey, considering the distinct nature and needs of every flock. Flock dynamics, including its size, age distribution, and breed, significantly influence the type of feed that best matches their needs. What ensures peak health for one coop may not necessarily suit another. Hence, maintaining sharp observational skills—attuning oneself to behavioral indicators like feather condition, activity levels, and egg quality—combined with making gradual adjustments based on chickens' responses to various feeds will guide you towards the most beneficial dietary balance, unlocking your flock's potential to sustain and thrive.

CHOOSING THE RIGHT FEED AND SUPPLEMENTS

Selecting the optimal feed is a nuanced process hinged significantly on the specific goals and compositions of your flock. For egg-laying hens, layer pellets emerge as the ideal nutritional blend. These pellets are crafted to combine ample protein with necessary calcium to support consistent egg production while maintaining overall health. The carefully calculated balance ensures hens remain in prime health, avoiding conditions such as osteoporosis. Conversely, broiler feed is apt for chickens raised for meat, possessing higher energy content to facilitate their rapid growth. This feed often includes a blend higher in corn and soybean meal—components renowned for their energy-yielding properties.

Understanding the nuances of a nutritional pyramid is crucial for choosing feed that aligns with your chicken-keeping goals. This pyramid, similar to those used in human nutrition, emphasizes the balanced integration of proteins, grains, minerals, and vitamins, each layer contributing to the overall health and productivity of your chickens. At its base, grains provide essential carbohydrates for energy, while the next layer up, proteins, support muscle and feather development. Above proteins, vitamins and minerals form a critical layer, ensuring immune health, strong eggshells, and robust skeletal structures. The apex of the pyramid focuses on supplements such as calcium and grit, enhancing diet specifics where standard feed may fall short. By familiarizing yourself with this structured approach, you can more effectively assess whether a specific feed formulation meets the comprehensive dietary requirements of your flock, ensuring they receive the balanced nutrition necessary for optimal health and egg production.

Location-specific availability of these feeds necessitates interaction with local farm stores or co-ops, where one can not only discover varied feeding options but also benefit from the guidance of fellow chicken enthusiasts. These resources offer invaluable insights and advice tailored to your locality's specific farming ecosystem. Engaging with the local farming community cultivates a network of shared knowledge, experiences, and collective purchasing power.

Supplements are essential dietary aids, acting as crucial bolsters where primary feed falls short of comprehensive nutrition. For instance, oyster shells serve as a vital calcium boost to bolster strong eggshell formation and combat issues like egg binding. Chickens are susceptible to a range of conditions if their diet is not adjusted for their specific physiological demands, making careful observation and strategic supplement incorporation essential. Likewise, grit is another necessary supplement, particularly for chickens lacking natural access to gritty soil, such as non-free-range birds. This gritty material aids in the grinding of food inside the gizzard, an essential process for digestion due to chickens' lack of teeth.

Seasonal changes and climate variables demand considerable dietary adjustments for chickens. When winter's icy grasp encroaches, their energy needs increase to sustain body warmth, prompting shifts in feeding strategy. Corn, with its high calorific content, becomes akin to a warming coat against winter's chill, stabilizing their metabolism amid frigid temperatures. Conversely, during the peak summer when the sun blazes with ferocity, maintaining hydration is paramount. Offering juicy fruits or hydration-rich vegetables like watermelon furnishes a welcome reprieve from the heat, helping chickens regulate body temperature while guaranteeing essential hydration levels. Such considerations ensure the flock's comfort and productivity are maintained regardless of climatic upheavals.

Understanding the cyclical nature of the chickens' breeding and laying patterns can further enhance the adjustment of their feed. During peak laying seasons, a higher concentration of calcium will be necessary, while during molting phases, increased protein and supplementary vitamins assist in feather regrowth and maintaining overall health.

Ensuring quality in chosen feed brands can significantly uplift poultry health and productivity. Brands such as Purina Layena and Nutrena NatureWise have gained recognition for their specialized formulations supporting poultry vitality. Purina Layena is precisely engineered for optimal egg production, while Nutrena NatureWise offers a range accommodating various poultry needs with its balanced formulations. These branded feeds are regularly tested to uphold high-quality standards, ensuring chickens receive adequate nutrition for maximal health benefits. Prioritizing these brands and maintaining consistency in feed quality ensures sustained health and productivity in your flock.

Reducing feeding costs without diminishing quality demands strategic planning and savviness. Purchasing in bulk offers discounts that accumulate into significant savings over time. Many local feed stores promote bulk purchases, occasionally accompanied by loyalty programs for additional cost benefits. Moreover, crafting homemade feed mixes tailored to your flock's requirements offers further opportunities for cost-saving, enabling you to purchase individual feed components economically and fashion a custom blend that meets nutritional requisites without relying heavily on commercial feed products.

Mitigating waste stands crucial for managing feeding costs effectively. Adequate storage solutions, such as airtight containers, ward off spoilage and pest infestation—a vital countermeasure considering most grains' susceptibility to these risks. Feed troughs designed to minimize spillage help conserve food, ensuring maximum nutrition reaches the chickens rather than being scattered and wasted. Implementing these storage practices prevents rodents and other pests, helping maintain nutritional integrity and lowering potential health risks.

Incorporating kitchen scraps provides an enriching nutritional variation while decreasing food waste. Foods like vegetable peelings, fruit scraps, and leftover grains add dietary diversity. Nevertheless, it is essential to avoid harmful foods such as avocados or onions. Removing uneaten leftovers promptly and chopping larger scraps prevent spoilage and deters pests, maintaining a clean, safe feeding environment.

Adjusting dietary plans for seasonal changes is imperative. Ensuring proteins are available for feather insulation during the colder months, while hydration-rich foods take precedence during summer's heatwaves, aligns diet with environmental demands. Reactive adjustments based on sudden climatic changes could involve increasing energy-rich foods during cold spells or offering cooling treats during heatwaves.

Ensuring clean and uninterrupted access to water is a non-negotiable requirement for maintaining flock health. Hydration is critical for digestion, egg production, and overall vitality. Automatic waterers and heated water bowls offer practical solutions to provide constant water supply even during freezing conditions, thus eliminating the need for frequent manual intervention during harsh weather conditions.

These strategies form the foundation of a year-round nutrition-focused approach to poultry care, establishing a sustainable path towards nurturing a thriving, robust flock. Consistency in application ensures the flock remains resilient and productive, aligned with scheduled husbandry cycles that mirror the natural environment chickens thrive in.

MANAGING FEEDING COSTS WITHOUT COMPROMISING QUALITY

Upon commencing their journey into chicken raising, many have been taken aback by how quickly feed expenses could accumulate. Establishing cost-effective practices should become a cornerstone of an efficient management strategy, ensuring the flock's health without exerting excessive financial pressure. The realization that small decisions accumulate to bear significant financial implications will inform one's drive towards cost-conscious feeding practices. For many, embracing bulk purchasing discounts have emerged as a primary strategy for substantial savings. Collaborating with local cooperatives or partnering with suppliers often yields opportunities for alluring bulk discounts. This strategy can be more effective when organizing with other local chicken keepers, allowing small-scale buyers to access bulk rates traditionally reserved for larger operations. However, these purchases necessitate verifying the adequacy of storage facilities to manage larger quantities effectively without encountering spoilage or infestation issues.

Enhancing dietary strategies through homemade feed blends is a beneficial practice embraced by experienced chicken keepers. This avenue permits tailoring feed to align precisely with nutritional needs, offering greater control and potentially boosting cost efficiency. Exploring diverse recipes that dovetail with chickens' nutritional mandates is advisable, offering flexibility in adapting feed compositions to suit seasonal or breed-specific needs. Opt for grains such as oats or barley as the foundational feed base, enriching it with protein fortification derived from fish or soybean meals. Such hands-on approach enables control over ingredient quality and makes nutritional customization accessible, reducing dependency on generic commercial products. This do-it-yourself approach provides transparency in ingredient selection and can economically align with nutritional aspirations comparably better than commercial feed reliance.

Regularly tracking your expenses for feed and supplements is key to managing your budget effectively. By using budgeting tools or templates, you can keep a close eye on your spending, offering a comprehensive overview of your costs. Recording your expenses for storage and maintenance supplies also helps reveal how resource use varies with the seasons, pinpointing opportunities for budget refinement or essential investments. This diligent financial tracking is crucial for optimizing your financial management, securing the long-term sustainability of your chicken-raising venture.

BUDGETING TEMPLATE

- **Monthly Feed Expenses**: Record bulk purchases and individual bags.
- **Supplement Costs**: Catalog additional items such as grit or oyster shells.
- **Storage Expenses**: Factor in costs for airtight storage containers and pest control measures.

- **Unexpected Costs**: Keep track of any additional unexpected expenses such as veterinary care or emergency feed solutions.

Utilizing these financial management templates is crucial for making well-informed decisions that prioritize cost-efficiency without skimping on your flock's nutrition. The core of effective budgeting is to sustain practices that ensure quality care without overspending, thereby promoting the well-being of your flock within a manageable budget. This strategy safeguards the health of your chickens while also streamlining expenses to reinforce the economic foundation of your chicken-keeping venture.

Building networks within local communities brings you closer to fellow chicken enthusiasts, fostering collaboration in bulk purchases or shared experiences. Such camaraderie nurtures a vibrant, innovative atmosphere that enriches your poultry-keeping journey, potentially leading to joint endeavors in feed production or shared success stories in devising cost-saving strategies. The knowledge exchange in local farming groups often highlights inventive solutions one might not have independently considered, offering sustainable ways to reduce costs without undermining the nutritional quality delivered to the flock.

Embracing these strategies not only conserves resources and enriches the chicken-raising experience but also bolsters flock care quality, devoid of unnecessary financial burdens. By leveraging collaborative knowledge, embracing strategic planning, and refining feed selection, nutritional needs can be met meticulously while respecting budgetary constraints, ensuring the flock—and its caretakers—prosper sustainably.

CHAPTER 4
HEALTH AND WELLNESS OF YOUR FLOCK

RECOGNIZING AND TREATING COMMON CHICKEN DISEASES

T he first time I observed one of my neighbor's cherished hens appearing unwell, it made a lasting impression on me, highlighting the deep care and responsibility involved in raising chickens. The signs of her ailment were unmistakable: her previously lively comb had become droopy, resembling a wilting flower, a clear sign she was not her usual spirited self. This incident prompted me to delve into the various diseases that can affect poultry, each presenting a unique threat to their health. Among these concerns, respiratory infections such as Infectious Bronchitis beckon immediate attention. This particular illness weaves through flocks with unnerving speed, characterized by symptoms such as persistent sneezing and a decline in egg production, which could spell economic loss for a farmer relying on egg sales. Its rapid spread underscores the necessity for early intervention.

Equally distressing are parasitic threats like coccidiosis, especially notorious among the avian community. Affecting primarily young chicks, coccidiosis manifests through diarrhea and stunted growth, emphasizing the criticality of vigilance from the onset of life. A proactive approach in detecting illness involves keen observation, honing skills to detect the slightest shifts—a despondent demeanor, avoidance of the flock, or a lack of interest in regular feed. These signs could be precursor signals of underlying health issues that demand prompt response.

Effective treatment is the backbone of flock health, providing relief and recovery when faced with adversities. Antibiotics, judiciously administered under strict veterinary supervision, can be pivotal against bacterial ailments. However, their careful and controlled use is critical to prevent undue resistance. Exploring complementary treatments, such as natural remedies, can serve as an adjunct to traditional medical care. Garlic, known for its mild antibacterial properties, can be introduced by crushing it into the water supply, potentially forestalling minor infections. Nonetheless, it's imperative to recognize when these home remedies fall short and professional medical intervention must take precedence, ensuring the provision of the best possible care to your flock.

Beyond treatment, veterinarians play a vital role in preventive health measures, guiding you to preserve the general well-being of your flock. Building rapport with a vet seasoned in poultry care could hold immense value, ensuring the availability of advice for varied situations. Investing time in crafting a health protocol tailored to your flock's unique needs, along with conducting regular veterinary screenings, can preempt many potential health issues.

ADDITIONAL SYMPTOM INDICATORS

These newly introduced symptoms could also offer a broader view:

Symptom	Possible Cause	Action
Lethargy	Viral Infection or Nutritional Deficit	Conduct blood tests, reassess diet, consult a vet
Sudden Weight Loss	Worm Infestation or Lowered Immunity	Initiate deworming, supplement vitamins, seek vet
Limping	Injury or Infectious Arthritis	Isolate and observe, provide anti-inflammatory care

Consistent monitoring reinforces your ability to swiftly identify disorders, forming the core of an effective response strategy. This knowledge empowers reaction, transforming potential health challenges into educational journeys, fostering a nimble, prepared approach toward chicken keeping.

EFFECTIVE PARASITE CONTROL AND PREVENTION

When first venturing into the realm of parasites, it can seem daunting; however, understanding these stealthy invaders is crucial to safeguarding your flock. Both internal and external parasites threaten chicken health, with lice being one of the smallest yet most troublesome. These minuscule insects may inhabit the feathers, causing discomfort and potential bald patches due to persistent scratching. Mites, elusive by nature, make their moves in the stillness of the night, eschewing daylight and latching onto their hosts only under the guise of darkness. Among internal parasites, notorious worms like roundworms and tapeworms wreak havoc on a chicken's internal system, siphoning essential nutrients and causing weight loss if untreated.

A strategic prevention plan is crucial to avoiding infestations altogether. The foundation begins with cleanliness; maintaining a tidy coop environment becomes your first line of defense. Employing diatomaceous earth—a powder derived from ancient deposits of fossilized algae—acts as a natural deterrent and disrupts parasite life cycles. Consider it akin to constructing a stronghold against invaders. Moreover, regular changes of bedding, as well as ensuring proper ventilation, create an inhospitable environment for these pests, keeping them at bay.

Addressing active parasite infestations necessitates a multifaceted approach to treatment. Utilizing topical solutions such as dusting powders and sprays can effectively target lice and mites, while a deworming schedule, as recommended by a veterinarian, ensures internal parasites are also managed. Consulting with poultry health experts can enhance your approach, offering proven strategies to safeguard your flock's health. Additionally, incorporating natural deterrents, such as oregano oil, offers a chemical-free method to repel pests. A few drops in their water can serve as an effective, nature-inspired preventative measure, blending the best of traditional and holistic parasite control techniques.

A finely honed checklist aids in the ongoing battle against these intruders:

ADVANCED PARASITE CONTROL CHECKLIST

- Implement rotational grazing to break parasite life cycles.
- Incorporate antiparasitic plants such as wormwood within run space.
- Designate ample dust bathing areas to promote natural cleaning.

Step	Action
Coop Maintenance	Weekly deep clean, replace or sanitize bedding
Birds Inspection	Inspect for signs of parasites bi-weekly, focusing on vents and feathers
Herbal Integrations	Plant herbs like thyme and lemon balm near the coop
Natural Treatments	Integrate feeds with natural antiparasitic properties

With vigilance and preparedness, even the tiniest of threats can be neutralized, maintaining your chickens' thriving condition. By combining thorough maintenance, proactive treatment, and preventive measures, you empower your flock with the tools needed to live vibrantly. A pest-free coop is a testament to diligence, enabling you to reclaim serenity knowing your efforts yield a robust and lively flock.

VACCINATION BASICS FOR A HEALTHY FLOCK

Initially, the concept of vaccinations was overwhelming and somewhat unclear. However, it quickly became evident that vaccinations are essential for safeguarding the flock's health. Vaccinations erect an essential barrier against formidable diseases. For instance, outbreaks of Newcastle disease, a highly contagious respiratory virus, can wreak havoc throughout a flock. Vaccinating chickens greatly diminishes its spread, so their overall health is sustained, which dovetails with the owner's peace of mind. Marek's disease is another complex challenge—known for causing tumors and paralysis without warning. Vaccinations aimed at this disease significantly curtail its reach, enhancing the flock's collective resilience.

Certain vaccines hold prominence in the backyard chicken realm. The Fowlpox vaccine stands as a bulwark against a virus notorious for grotesque lesions and formidable health detriments. Similarly, vaccinating against Avian Encephalomyelitis shelters young chicks from a viral disease leading to tremors and motor dysfunction. Incorporating these vaccinations into your flock's health plan efficiently shields against potential catastrophe, nurturing a healthy, thriving group of chickens.

Timing is a critical element in the vaccine regimen. Initial vaccination doses are typically administered to chicks when they're only a few days old to establish foundational immunity. Booster schedules are essential to maintaining continuous protection against evolving pathogenic threats. Regularly monitoring these schedules and adhering to follow-up doses helps achieve optimal long-term efficacy.

Multiple vaccination methods afford flexibility based on specific requirements. Injectable vaccines deliver precise dosages, albeit necessitating the handling of each individual bird. Alternatively, water-based vaccines offer a less invasive option, ideal for larger flocks, although they require exact mixing and meticulous application to ensure thorough dissemination.

Compliance with vaccine packaging instructions is essential for effectiveness and safety, necessitating proper storage and handling. These guidelines maximize vaccine potential, ensuring it remains a robust tool in maintaining flock health.

Vaccinations embody a strategic investment to bolster productivity and diminish worries over poultry health, preserving peace of mind and yielding prosperous chicken endeavors.

Sourcing vaccines might often be challenging for smaller operations as commercial solutions cater predominantly to large-scale enterprises. Acquiring pre-vaccinated birds through hatcheries circumvents logistical concerns, providing a seamless and effective alternative.

While integrating vaccinations within your management plan is crucial, they are just one tier of comprehensive flock care. Paired with stringent hygiene practices and routine health audits, vaccinations become part of a holistic strategy leading to optimal poultry well-being.

Each proactive step warrants an environment conducive to thriving chickens. Armed with a vaccine regimen, you will be equipped to robustly defend against potential challenges, cultivating a vibrant and dynamic flock.

UNDERSTANDING CHICKEN BEHAVIOR AND SOCIAL DYNAMICS

Chickens inhabit a world characterized by richness in social dynamics. Chickens establish a social hierarchy, commonly known as the "pecking order," that is crucial for flock harmony. This system determines access to vital resources such as food and preferred roosting spots. Seeming from the outside like a cacophony, this precise order is fundamental in maintaining intra-flock harmony. The introduction of new hens disrupts this balance, necessitating rank establishments and triggering necessary scuffles as each bird finds its footing. Although this may seem aggressive, it's a natural integration process that stabilizes once order restores within the hierarchy.

Foraging and dust bathing, more than mere routine activities, are essential for chicken satisfaction. Foraging encompasses more than aimless pecking; it's an exploratory activity tapping into their instincts while offering mental and physical stimulation. Dust bathing operates as a form of self-cleaning, an effective means to combat parasites, and a source of evident joy. Watching them whirl in delight within the dust privileges the onlooker with a glimpse of their innate happiness and contributes to health preservation.

The presence of a dominant hen or rooster sheds light on the intricacies governing social dynamics. Often more assertive rather than the largest, this bird naturally assumes leadership, guiding the flock steadily between resources and shelter, fostering a sense of tranquility. Nonetheless, vigilant monitoring is crucial to intercept aggression that sometimes ensues from excessive authority. Instances of bullying must be addressed promptly, ensuring harmony and welfare across the flock spectrum.

Sensitivity to behavioral fluctuations, such as sudden withdrawal, often signals underlying discomfort. Environmental precursors, such as changes in weather, often dictate shifts in behavior patterns, serving as vital predictive cues. Strategic redirection via new perches or introducing hanging vegetables as interactive distractions assists in curbing undesired conduct. Creating an environment filled with engaging activities reduces behaviors caused by boredom, leading to healthier interactions among the flock.

Keeping a detailed behavior journal aids in elevating understanding and documentation of flock interactions:

BEHAVIOR JOURNAL PROMPT

- Daily interactions and observations.
- Noting any behavioral shifts or changes in dynamics.
- Recording intervention attempts and corresponding results.

Maintaining a detailed record offers valuable insights and serves as a resourceful guide for fine-tuning social dynamics approaches. Enhanced comprehension of chicken behavior solidifies your management repertoire while intensifying your intrinsic connection with the flock.

MANAGING STRESS AND ENSURING FLOCK HARMONY

Chickens, like all beings, endure stress, impacting both health and productivity. Spotting potential stress inducers is integral to promoting a serene living environment. Overcrowding, a prominent catalyst, incites competition for essentials such as food and water, drawing parallels to our own discomforts amidst restricted resources. Ensuring ample space both within the coop and the run alleviates such pressures considerably. Additionally, abrupt environmental shifts, such as relocating the coop or altering feeding schedules, can disturb chickens, disrupting the tranquility of their environment.

Mitigating stress requires thoughtful strategies, with gradual introductions of new members easing potential tension. Allowing visual contact prior to cohabitation smooths subsequent transitions. Reliable care routines contribute to flock security by providing predictability and trust, dissolving uncertainties tied with unpredictability.

Fostering harmony necessitates methodology. Installing

multiple feeders reduces resource competition akin to distributing buffet stations, enhancing accessibility during crowded gatherings. Providing plentiful roosting space ensures peaceful rest, mitigating conflicts over prime spots. Proper arrangements promise comfort for all birds, reducing disputes substantially.

Implementing minor adjustments translates into significant strides towards flock peace. During the integration of newcomers, initial chaos might appear formidable, yet strategic interventions like temporary barriers or adjustments to yard access encourage gradual reconciliation, attaining harmony over time. Expanding physical space with portable fencing reframes dynamics, alleviating pecking disputes resulting from spatial constraints.

In addition to physical adjustments, mental stimulation plays a crucial role in maintaining flock harmony. Introducing activities that engage their minds and bodies significantly reduces aggressive behavior, leading to a more peaceful group. Providing hanging vegetables or creating designated dust bathing areas adds enrichment to their daily routine, akin to the comfort toys offer during unfavorable weather conditions. A mentally engaged and happy flock is often more productive, enhancing the overall health and harmony within the coop.

Managing stress and maintaining harmony within your flock requires a delicate equilibrium. Yet, each improvement you make significantly contributes to the tranquility of your chickens. By observing their interactions closely, you can tailor strategies that specifically cater to their needs, fostering a peaceful environment. Embracing flexible approaches and being willing to adjust your methods encourages a harmonious balance, ensuring your flock thrives in a stress-free setting.

To create a tranquil haven for your chickens, it's essential to grasp and cater to their needs thoughtfully and deliberately. Recognizing and reducing stress factors, coupled with nurturing a harmonious environment, builds a strong foundation for a thriving chicken community.

HOLISTIC APPROACHES TO CHICKEN HEALTH

Holistic health adds a layer of vitality to chicken care, prioritizing their overall well-being to enhance resilience and energy. Incorporating herbal supplements like echinacea and elderberry into their feed or water leverages their immune-boosting benefits.

Similarly, incorporating probiotics plays a crucial role in fortifying gastrointestinal health by introducing a diverse community of beneficial bacteria. This introduction supports a more balanced digestive ecosystem, facilitating enhanced nutrient absorption and bolstering the immune system. Available in forms such as powders, liquids, or directly infused into feed, these supplements can be easily integrated into your chickens' daily diet. The consistent use of probiotics not only contributes to improved digestive health but also promotes overall well-being, leading to happier, more content chickens. By prioritizing gut health through the strategic use of probiotics, you're investing in a foundational aspect of your flock's holistic care, ensuring they maintain optimal health and contentment.

Cultivating a healthy environment remains essential. Sunlight exposure facilitates vitamin D synthesis, invaluable for bone strength and robust egg production. Strategically positioning coops or incorporating windows maximizes sunlight access. Alongside light, herbs like mint and lavender exude aromatic charm while concurrently repelling insects, crafting an inviting habitat.

Promoting natural foraging behaviors allows chickens to enhance their diet beyond what standard feed offers. Their inherent curiosity drives them to peck, scratch, and graze, uncovering a variety of nutrients not found in their usual meals. This engagement with their environment not only fulfills their nutritional needs through a broader spectrum of resources but also activates their natural instincts, adding richness to their daily routines.

Balancing holistic approaches with traditional care methods integrates a comprehensive health strategy. Genuine symbiosis between vaccinations and natural remedies offers a reliable defense, fused with enriched resilience.

Embracing holistic care encompasses a blend of preventative and curative approaches, setting the foundation for a thriving flock. Picture your chickens basking in the warmth of the sun, their feathers shimmering with health, embodying the essence of vitality. This approach nurtures an environment that allows your chickens to thrive, promoting their well-being and ensuring a life of vigor and longevity.

In conclusion, holistic health transcends addressing surface challenges, instilling environmental support fostering physical and mental wellness. The next chapter will delve deeper into the egg production process, journeying from coop life to kitchen transformations.

MAKE A DIFFERENCE WITH YOUR REVIEW

UNLOCK THE POWER OF GENEROSITY

"The best way to find yourself is to lose yourself in the service of others." – Mahatma Gandhi

People who give without expecting anything back often feel the happiest. So let's do something kind together!

Would you help someone just like you—curious about raising chickens but unsure where to start?

My goal is to make raising chickens simple, fun, and rewarding for everyone.

But to reach more people, I need a little help.

Most folks choose books based on reviews. So I'm asking you to help a fellow chicken-keeper by leaving a quick review.

It's totally free and takes just a minute—but it could make a big difference in someone's chicken journey. Your review might help…

…one more family gather fresh eggs from their backyard.

…one more kid learn where food comes from.

…one more person enjoy the peace that comes from caring for animals.

…one more beginner feel confident to start.

…one more dream of a simple, happy life come true.

To leave a review, just scan the QR code or visit this link:
https://www.amazon.com/review/review-your-purchases/?asin=B0FD478RY5

If you enjoy helping others, you're my kind of person.

Thank you so much for your support—it truly means the world.

— Avery Sage

CHAPTER 5
EGG PRODUCTION AND QUALITY

FACTORS INFLUENCING EGG PRODUCTION IN CHICKENS

E nvision the comfort of your morning routine, cradling a warm cup of coffee as you admire a basket filled with fresh eggs gathered from your own backyard hens. This moment of tranquility highlights the deep bond shared between you and your flock. To sustain these quiet moments, understanding the myriad factors influencing egg production is crucial. For example, environmental conditions play a pivotal role. As daylight wanes with the onset of winter, egg production tends to decline. Chickens thrive on 14 to 16 hours of light a day to sustain peak laying performance. To counteract nature's tempered light during darker months, think about integrating supplemental lighting. Employing artificial lighting wisely can recreate the effect of longer daylight hours, encouraging your flock to sustain their egg-laying rates even through the shorter days of winter.

However, not just artificial lighting supports productivity; the strategic placement of windows in the coop can harness natural light, creating a more environmentally conscious approach to augmenting light exposure. This dual approach blends the benefits of technology with nature's rhythms, fostering an environment where hens feel aligned with natural cycles despite seasonal changes.

Temperature fluctuations rank high among variables affecting chicken productivity. In sweltering heat, hens expend vital energy regulating body temperature, while in biting cold, their efforts shift toward generating warmth over egg production. It's essential to create a controlled environment that minimizes these extremes. Implementing insulated coops, installing fans for ventilation, and erecting shade structures can regulate temperatures favorably, ensuring your hens remain comfortable, thus optimizing their laying cycles.

Beyond these solutions, the importance of microclimates within a coop could be explored further. By positioning water sources and ventilation openings strategically, multiple temperature zones can be established within a single coop. This variety offers hens choices, enabling them to seek out the most comfortable conditions to maximize their productivity.

Genetics is another cornerstone influencing output. Egg-laying breeds, like the highly efficient Leghorns, showcase remarkable genetic capabilities, yielding upwards of 300 eggs each year. Your selection of breeds inevitably decides the yield—heritage breeds like Orpingtons or Sussex may produce fewer eggs annually but offer exceptional qualities like sturdiness and amiable temperaments. A comprehensive understanding of your breed's genetic makeup guides realistic expectations, ensuring that goals align with biological possibilities.

Expanding upon breed considerations, the genetic inclination towards winter laying is an additional aspect worth examining. Certain breeds inherently manage colder conditions better, continuing to lay throughout the winter with minimal encouragement. This genetic insight can help orient new flock introductions based on regional climatic patterns—choosing a breed that inherently fits your environment maximizes success.

Age is a significant determinant of egg production. The prime years, typically between one and two years old, showcase a hen's peak laying potential, often leading to abundant egg collection. Contrarily, as hens age beyond three years, egg production diminishes progressively, a shift reflecting their physiological tilt towards sustenance and corporeal maintenance over reproduction. Strategically planning the timeline for introducing new birds into your flock, therefore, plays a key role in ensuring a continuous and balanced supply of eggs.

Even more, incorporating temporary substitutions into the flock, such as younger hens or pullets, can mitigate the natural production dip as older hens taper off. This consideration ensures that production levels remain steady, as the younger birds begin to compensate for the reduced output from their elder counterparts.

Stress and health are substantial influencers on egg yield. Frequent predator sightings often disrupt your flock's routine, potentially halting their laying. Guaranteeing a secure coop and run is vital for shielding chickens, reducing fear-induced production disruptions.

Health complications, such as egg binding—a serious condition where an egg becomes stuck within the hen's reproductive tract— demand prompt and careful attention to avert critical health issues. Egg binding can be caused by a variety of factors, including nutritional deficiencies, particularly a lack of calcium, obesity, or genetic predispositions. Recognizing the signs of this condition early, such as a hen straining without producing an egg, lethargy, or a swollen abdomen, is crucial for timely intervention. Immediate measures might include gently warming the hen to relax her muscles or consulting a veterinarian for professional advice. Preventative strategies, focusing on proper nutrition, regular exercise, and stress reduction, play a key role in minimizing the risk of egg binding and ensuring the health and productivity of your flock.

INTERACTIVE ELEMENT: EGG PRODUCTION JOURNAL

Consider launching an egg production journal. Document environmental factors, daylight variations, and stressors impacting your flock. Regular entries serve as diagnostic tools, unveiling patterns and offering insights into laying trends. Proactive management through journaling enables you to make informed decisions, enhancing egg production. Recognizing alterations in laying patterns early can illuminate the need for interventions, fostering adjustments that optimize productivity. This simple yet powerful practice acts as a guide for sustaining balance within the flock, ensuring that each egg collected mirrors the efforts invested in creating an ideal living environment for your hens.

Incorporating various tracks, such as egg size and weight over time, can further enrich this journal, offering comprehensive insight into both immediate conditions and longer-term trends. The journal not only acts as a day-to-day log but evolves into a valuable resource for fine-tuning hen management strategies.

ENHANCING EGG QUALITY WITH PROPER NUTRITION

Cracking open an egg to reveal a vibrant yolk housed in a shell of pristine firmness speaks volumes of the nutritional care provided to your chickens. Essential nutrients play pivotal roles in achieving such quality, beginning with calcium—indispensable for forming strong shells. Insufficient calcium intake often results in fragile, easily cracked shells. Offering calcium supplements like crushed oyster shells or limestone fortifies shell structure, creating a preventive barrier against breakage. Omega-3 fatty acids are equally vital for augmenting yolk quality. Common in ingredients like flaxseed or fish meal, these beneficial fats enrich the nutritional value of your eggs, enhancing both their visual appeal and health benefits.

Moreover, vitamin D is crucial in calcium absorption, and ensuring your flock has access to natural sunlight can exponentially increase egg quality. For coops situated in areas with limited sunlight exposure, vitamin D supplements can fill the gap, sustaining the interdependent relationship between calcium structures and vitamin effectiveness.

A balanced diet is crucial for maintaining exceptional egg quality. Layer pellets, rich in proteins, vitamins, and essential minerals, should form the foundation of this diet, tailored specifically for laying hens. The monotony of pellets can be expanded upon by integrating fresh greens like kale or spinach. These additions not only diversify dietary intake but also infuse eggs with enhanced color and flavor, ensuring a broad spectrum of nutrients reaches your flock. Such variety promotes robust health, enhancing productivity.

Natural additives present another layer of nutritional fortification, readily elevating egg quality. Flaxseed, abundant in omega-3s, boosts both yolk color and content. Incorporating flaxseed into the diet manifests visually and health-wise, producing golden yolks that stand as testaments to nutritional diligence. Apple cider vinegar, an unsung hero in poultry nutrition, aids digestion and fortifies gut health. Adding a splash to your hens' water improves overall nutrient absorption efficiency, translating into tangible benefits reflected in egg quality.

Dive deeper into the idea of homemade feed options—an innovative approach that satisfies specific dietary needs while minimizing food waste. Scraps such as citrus peels and small amounts of garlic can offer antioxidative benefits, subtly complementing standard poultry feeds. This judicious practice reduces waste and contributes to a closed-loop system in backyard husbandry.

Crafting homemade feed is an alternative that meets specific dietary needs while reducing waste. Kitchen scraps find renewed purpose in this approach—remnants like carrot tops and lettuce leaves become nutritious supplements when chopped and mixed with grains such as corn or oats. This practice transforms potential waste into fresh dietary additions, supporting your flock's health.

Custom feed mixes offer further tailoring opportunities. Begin with base grains—corn or wheat blends well—supplementing with protein-rich ingredients like soybean or fish meal. Seeds add nutritional and entertainment value, enriching your flock's diet. This approach empowers you to dictate precise nutritional inputs, maximizing health and egg quality potential.

HOMEMADE FEED RECIPE

A straightforward recipe might involve:

1. **Base Grains:** 50% corn, 20% wheat
2. **Proteins:** 10% soybean meal, 10% fish meal
3. **Additives:** 5% flaxseed for omega-3 enrichment, 5% sunflower seeds for taste and texture

Thoroughly mix these elements and store in airtight containers to preserve freshness. Experiment with portion sizes to accommodate fluctuating seasonal needs, understanding that dietary tweaks might be necessary to align with changing production phases.

Balancing dietary needs while enhancing egg quality invites creativity into chicken husbandry. A diet rich in calcium and omega-3s fortifies eggshells and imbues yolks with vibrance. Supplementing with fresh greens and natural additives nurtures overall health and laying proficiency. Homemade feeds utilizing kitchen scraps not only minimize waste but also grant control over nutritional dynamics. Adjusting these diets seasonally or based on specific needs ensures chickens remain healthy and productive year-round. This careful curation transforms basic ingredients into enriching meals, culminating in eggs of superior quality—a nourishing delight for both chickens and their keepers.

TROUBLESHOOTING COMMON EGG PRODUCTION ISSUES

The disappointment of fewer-than-usual eggs or eggs with unusual shapes and textures is a shared frustration among chicken keepers. Diagnosing the core issues requires a blend of observation and informed detective work. Nutritional deficits often lie at the heart of egg production challenges. Inadequate calcium intake, for example, manifests in thin or soft eggshells. Similarly, a deficiency in essential nutrients can lead to a downturn in egg count.

Delving deeper into the aspect of feed palatability, we uncover additional complexities. Changes in the flavor profile of feed, whether from ingredient degradation or shifts in the quality of feed components, can impact hens' willingness to eat adequately. This reluctance can negatively influence egg production. By conducting regular taste tests of feed batches, any issues with feed attractiveness can be identified early, allowing for swift corrective measures.

Mitigating stress factors is essential for maintaining egg production. The presence of predators or sudden changes in their environment can cause hens to stop laying as part of a natural survival response. To counteract this, reinforcing coop security, providing a diet rich in essential nutrients, and effectively managing any stressors are key measures.

Real-life experiences provide invaluable insights. Consider a seasoned chicken keeper who faced a calcium deficiency in her birds, resulting in fragile shells. By augmenting their diet with oyster shell supplements and ensuring a nutritionally balanced feed, production levels improved remarkably within weeks. Another scenario involved a drop in egg numbers following a hawk sighting. Installing reflective deterrents and providing enclosed shelters restored the flock's comfort and laying patterns, illustrating how addressing stressors impacts productivity.

Understanding your flock's needs, observing behavioral shifts, and implementing targeted solutions form the core of effective egg production troubleshooting. Each issue resolved adds to your skill set, enhancing your flock's well-being. Although challenges exist, the satisfaction gleaned from overcoming them is immense, reinforcing the value of care invested in chicken keeping.

EFFICIENT EGG COLLECTION AND STORAGE TECHNIQUES

Walking into your coop to find a nest of pristine eggs is a delightful testament to your hens' labor. Effective egg collection and storage not only maximizes freshness but enhances enjoyment. Aim to collect eggs daily, preferably in the morning, when hens complete their laying. Regular collection mitigates spoilage risks and keeps eggs clean, deterring pests and ensuring quality.

Handling eggs with care is essential to avoid cracks and extend their shelf life. After collection, the focus shifts to proper storage to ensure they remain fresh. Storing eggs at approximately 45°F is ideal for maintaining their quality. While refrigeration is the most effective method, storing eggs at a cool room temperature, away from direct sunlight and heat sources, can also preserve their freshness. Placing eggs in cartons with the pointed end down helps maintain the yolk's central position, keeping the egg's structure and flavor intact.

Exploring methods to verify egg freshness is equally important for extending their utility. The float test is a simple yet effective technique that identifies the accumulation of gases in eggs, indicative of aging. By placing an egg in water, its buoyancy reveals its freshness—fresh eggs sink, while older ones may float. This rapid assessment tool helps in reducing waste, allowing for the early identification of eggs that are past their prime.

Traditional preservation techniques, such as water glassing, offer a nod to the age-old methods of extending egg freshness without the need for refrigeration. By submerging eggs in a solution of lime water, a protective barrier against bacteria is formed, significantly prolonging the eggs' shelf life. When stored in this manner, eggs can remain fresh for up to three weeks, even when kept outside the fridge. Delving into these historical preservation methods connects us to the innovative approaches of past poultry keepers, who skillfully navigated the challenges of egg storage long before the advent of modern refrigeration.

Exploring modern preservation techniques, like freezing eggs after they are cracked and stored properly, extends their lifespan significantly. This approach enables efficient inventory control, guaranteeing that no egg is wasted during times of high production when the rate of laying surpasses immediate consumption demands.

Whether rooted in ancestral lime water baths or facilitated by contemporary refrigerators, ensuring egg freshness reflects a commitment to quality. Appreciating every egg laid, knowing each was collected and stored with intention, uplifts the broader experience of poultry husbandry. Effective handling cultivates a sense of connection to the food cycle, affirming the bountiful returns derived from disciplined practices.

UNDERSTANDING THE MOLTING PROCESS AND ITS IMPACT

The sight of feathers strewn across your backyard may alarm initially, but it signifies the natural molting process. Molting is a regenerative cycle where chickens shed older feathers for new ones, impacting egg production. During molting, hens redirect energy towards feather growth, temporarily pausing egg-laying. Their bedraggled appearance belies the dynamic changes occurring beneath, as they prepare for renewal.

For some enthusiastic poultry keepers, understanding the biological nuances behind molting adds depth to compassionate care. Distinguishing between partial and full molts can alter your flock's management strategy. Partial molts, often minor and less demanding, contrast substantially with the extensive demands of a full molt, necessitating variable dietary adjustments.

A nutritionally demanding phase, molting calls for increased protein intake, akin to athletes in training. High-protein feeds, accented by nutritious snacks such as mealworms, support regrowth. Vitamins like A and E also contribute to recovery, promoting tissue health.

Behavioral changes accompany molting, too. Reduced activity levels allow hens to conserve energy. Temporary shifts in flock hierarchy are common as some birds, vulnerable due to feather loss, maneuver social structures. A caring approach involves adjusting the environment to meet nutritional and comfort requirements, facilitating a smoother transition.

Beyond addressing the direct energy needs of molting, maintaining an enriched environment during this transitional phase provides distractions, minimizing stress and subsequent social disturbances within the flock. Incorporating perches, dust bathing areas, and multipurpose foraging opportunities can enrich a hen's daily routine, preventing unnecessary angst.

Through this transformative process, patience and understanding become valuable assets. Witnessing the cycle of loss and renewal heightens appreciation for the resilience inherent in chicken life. By accommodating your flock's temporary needs, you enhance their health and strengthen the bond between keeper and hen.

ENCOURAGING LAYING WITH OPTIMAL COOP CONDITIONS

Creating an optimal coop environment is akin to curating a comfortable sanctuary where hens can lay their eggs freely. Proper nest box placement encourages privacy and tranquility, key to initiating and maintaining productive laying cycles. Position these boxes in serene areas, away from bustling activity to promote a peaceful laying atmosphere. Adequate space, lined with soft bedding, provides comfort and makes eggs less prone to breakage.

Building on this concept, introducing visual diversity in nesting areas—such as varied textures or gentle color contrasts—can create a comforting ambiance for laying hens. Keeping a clean environment while integrating these visual elements can encourage even the most reserved hens to utilize the nesting boxes, ensuring efficient use of all available space.

Consistent cleanliness promotes health and productivity. Regularly maintaining nesting and coop cleanliness deters pests and disease, reflecting care that engenders trust and freedom in hens' laying routines.

Lighting holds sway over egg production, with longer daylight encouraging regular laying. Shorter winter days necessitate supplemental lighting solutions, ideally set on timers to mimic natural patterns. By ensuring 14 to 16 hours of light daily, hens receive cues that spur consistent laying, independent of seasonal changes.

Reducing noise and managing stress are crucial for encouraging egg laying. A quiet and serene coop environment is vital for minimizing distractions and disturbances. Paying attention to the social interactions within your flock and resolving any conflicts can decrease stress levels, creating a harmonious atmosphere that supports consistent egg production.

Engage creatively by integrating sound control practices. Utilizing sound panels or introducing mild soothing background noise, such as recorded chirps or subtle classical music, can foster favorable auditory environments, reducing potential disturbances from external noise sources.

Enhancing coop conditions involves a mosaic of small adjustments that collectively promote a productive environment. From the strategic placement of nest boxes to implementing lighting and ensuring cleanliness, these measures nurture your flock's well-being while optimizing egg production. With each hen comfortably laying, the serene rhythm of chicken life continues amidst a backdrop of care and attention, setting the stage for bountiful harvests day after day.

In closing this exploration into egg production and quality, the stage is set for Chapter 6: Seasonal Care and Adaptation, where the changing seasons bring fresh insights and strategies for maintaining your flock's health and output year-round. With knowledge and practical application, each egg collected is a triumph, encapsulating the harmonious balance achieved between nature and nurture.

CHAPTER 6
SEASONAL CARE AND ADAPTATION

PREPARING YOUR FLOCK FOR WINTER WEATHER

As the first frost nips at your nose, you might be sipping hot cocoa, feeling snug in your wool socks, and pondering how on earth your chickens will fare in the chilly months ahead. Winter presents a unique set of challenges for backyard chicken keepers. But fret not! With a bit of preparation, your feathered friends can enjoy the cold season comfortably. Let's delve into how to keep them cozy and content as the mercury plummets.

Insulation is key to keeping your coop toasty. Think of it as swaddling your chickens' home in a warm, protective blanket. Installing insulation in the coop walls is a great starting point. Recycled materials like foam boards or even old carpets can be incredibly effective, acting as barriers against biting winds. If you're looking for more sustainable options, consider sheep's wool or recycled denim insulation, which offer excellent thermal resistance and are environmentally friendly. For added protection, consider using straw bales around the exterior of your coop. They not only provide excellent insulation but also serve as a windbreak, shielding your flock from those icy gusts that seem to slice right through you.

To enhance the insulation further, think about the coop's roof. A well-insulated roof prevents heat from escaping, and using materials like tar paper can seal in warmth effectively. Additionally, ensuring there are no gaps or leaks in the coop will help maintain a constant and comfortable temperature. Conducting a thorough inspection before the first snowfall can save a lot of hassle later on. Make sure to seal any possible entry points for drafts, as even a small gap can cause significant discomfort for your chickens.

Water management becomes a critical task during winter. The last thing you want is for your chickens' water supply to freeze solid. Heated waterers or electric heaters can be a godsend, ensuring that fresh water is always available. If electricity isn't an option, insulating your water containers with covers can help. A reflective cover with an insulating layer can prevent freezing, keeping the water at a drinkable temperature even when it's frosty outside. Using rubber buckets can also be beneficial as they retain heat better than metal or plastic versions, providing your chickens with an unfrozen drink as they navigate the chilly weather.

Innovative solutions are constantly emerging for keeping water thawed. Consider solar-powered water heaters, which can harness even the feeblest winter sunlight to keep water in a liquid state. Additionally, exploring the option of building thermos-like insulated containers for water storage can thwart the freezing elements. Even adding a few ping pong balls to waterers can prevent surface ice from forming due to their constant motion stirred by the wind.

Dietary modifications for winter are akin to fortifying your chickens against the cold. Increasing their calorie intake with cracked corn provides the extra energy they need to stay warm. Think of it as giving them an extra slice of pie at Thanksgiving. Additionally, foods high in fat and protein, such as sunflower seeds or kitchen scraps like oatmeal, can be introduced to give them an energy boost. Warm mash meals also serve as a treat, especially on those bone-chilling mornings. A mix of feed with warm water not only warms their bellies but also encourages them to eat more.

In the winter, when the ground is frozen and fresh greens are hard to come by, bringing the garden indoors can make a significant difference in your chickens' diet and health. Implementing a hydroponic system allows you to grow nutrient-rich microgreens, such as alfalfa, clover, and radish, right inside your home or a suitable outbuilding. This method of growing fodder bypasses the need for soil, using water enriched with minerals instead, and can yield a fresh supply of greens in just a matter of days. Including these microgreens in your chickens' diet not only introduces essential vitamins and minerals that may be missing during the colder months but also provides them with a varied diet, promoting better health and egg production.

Providing proper shelter and bedding ensures your chickens remain snug and dry. The deep litter method is an excellent way to generate warmth within the coop. By allowing bedding to accumulate and decompose, natural heat is produced, similar to a compost pile. Regularly adding fresh layers of straw or wood shavings will keep things clean and cozy. Ensuring dry bedding is crucial to prevent frostbite on those delicate chicken toes. Dampness is the enemy, so check often and replace bedding as needed. Consider using pine needles or shredded leaves for added insulation and a pleasant scent that can double up as natural mite repellents.

To supplement the deep litter method, explore using heat lamps cautiously. While they can provide targeted warmth, they also pose a fire risk if not positioned correctly. Consider using ceramic heat emitters that provide warmth without light, reducing stress while ensuring safety. Always integrate these systems with thermostat controls to maintain optimal temperatures.

WINTER READY CHECKLIST

- **Insulation**: Ensure coop walls are insulated; add straw bales or consider alternatives like sheep wool for wind protection.
- **Water Management**: Use heated waterers or insulated covers for water containers.
- **Diet**: Increase calories with cracked corn; offer warm mash, sunflower seeds, or oatmeal meals.
- **Shelter**: Implement the deep litter method; maintain dry bedding with additional layers like pine needles for warmth.

This checklist serves as a handy guide to keep your flock warm and comfortable despite the cold months. By addressing these aspects of winter care, you'll provide a secure, inviting environment where your chickens can thrive, despite the cold. And while winter might bring challenges, it also brings opportunities to deepen your connection with your flock as you work together to navigate the chill.

KEEPING CHICKENS COOL DURING HOT SUMMER MONTHS

Imagine a sun-drenched afternoon, the air thick with heat, and your flock looking slightly less perky than usual. Chickens, much like us, don't appreciate sweltering days. The good news is you can create a refreshing oasis for them with a few thoughtful adjustments. Start with promoting shade and ventilation. Installing fans or misters in the coop can work wonders to keep the air circulating. These aren't just luxuries—they're lifesavers when temperatures soar. If installing technology feels daunting, creating shaded areas with simple tarps or strategically planted trees can offer relief from the sun's relentless rays. Your chickens will thank you with every contented cluck.

Beyond artificial shade options, explore planting fast-growing trees that lose their leaves in winter but provide ample shade in summer, such as willows or poplars, strategically placed around the coop. This not only offers year-round shade benefits but will also enhance the coop's landscape aesthetics.

Hydration is another cornerstone of summer chicken care. Water isn't just about quenching thirst; it's essential for regulating body temperature. To ensure your chickens stay hydrated and healthy, especially during the warmer months, consider incorporating electrolytes into their water supply. Electrolytes are vital for maintaining a chicken's fluid balance and can help replenish the essential minerals they lose through exertion in the heat. This simple addition to their water can significantly impact their overall well-being. It's akin to sipping an ice-cold sports drink after a sweaty jog.

Frozen treats like watermelon slices can also provide a refreshing break and add a bit of excitement to their day. Picture your flock pecking eagerly at chilled fruity goodness—a delightful scene that keeps them hydrated and happy. Ice blocks with embedded fruits or vegetables can provide an engaging way for your flock to cool off while having some nutritious fun.

An alternative hydration method involves the use of nipple waterers or automated drinkers that ensure a constant supply of fresh, cool water. Consider installing a water filtration system to provide the purest refreshment, minimizing the risk of bacteria which often proliferate in heat.

Diet is another area where small tweaks can make a big difference. During summer, reducing corn intake is wise, as corn generates internal heat during digestion. Opt instead for high-water-content foods. Cucumbers, lettuce, and zucchini are excellent choices that provide hydration while being easy on their digestion. Think of it as swapping out hot soup for a crisp salad on a summer day—it just makes sense. Cold grains like barley can also be introduced, offering a nutritious and cooling alternative to the usual mix, ensuring that your chickens have the energy necessary without the added heat.

Moreover, the introduction of live insects or high-protein snacks ensures sustained energy levels for active summer days. Engaging your chickens with a diet rich in variety can further reduce stress and encourage healthy behavior.

Observing your chickens closely is key to catching signs of heat stress early. Panting, lethargy, and holding wings away from the body are chicken ways of telling you it's too darn hot. They might even look like they're practicing yoga with their wings stretched out like that! Recognizing these signs lets you step in before any real harm occurs. Strategically placed shade cloths or low-lying plants can also encourage your flock to find the cooler parts of the yard to retreat to in times of extreme heat.

Integrating heat-tolerant foraging plants or grasses into your run area can provide natural retreats and even nutritional benefits. Purslane and other drought-resistant plants can be grown to offer a cool respite and a source of omega-3-enriched greens.

SUMMER HEAT CHECKLIST

- **Shade & Ventilation**: Install fans/misters; create shaded spots with tarps, trees, or vegetation.
- **Hydration**: Add electrolytes to water; provide frozen treats like watermelon or ice blocks with embedded goodies.
- **Diet Adjustments**: Reduce corn; offer high-water-content foods and consider cold grains like barley.

- **Behavior Observation**: Look for panting, lethargy, or wings held away from the body, and encourage rest in cooler areas.

This checklist serves as a quick reference to keep your chickens comfortable and healthy throughout those sizzling summer months. The beauty of these strategies is that they're simple yet effective, allowing you to enjoy summer knowing your flock is well cared for. By addressing the needs of your chickens before the heat intensifies, you ensure that they remain cool, calm, and clucking contentedly all through the season.

ADJUSTING CARE ROUTINES FOR SPRING AND FALL

As the first green shoots of spring poke through the soil, it's time to adjust the care routines for your flock. Spring and fall are transitional seasons, each with their quirks and challenges. Both demand flexibility and awareness. One of the first tasks is gradually adjusting light exposure. Chickens are sensitive to daylight changes, which can affect their laying patterns and behavior. As days lengthen in spring, you might notice a boost in egg production, with hens busily scurrying to settle in their nests. In contrast, fall brings shorter days, signaling your hens to slow down. To help with this transition, gradually increase or decrease artificial lighting in the coop, mimicking natural daylight changes.

Fine-tuning artificial light in the coop not only aids in egg production but helps provide a consistent sleep cycle which can reduce stress. Opt for a timer-based lighting system to emulate natural crepuscular light gradients which can have calming effects on your flock.

Weather can be unpredictable during these seasons, with temperatures swinging wildly from warm to chilly in a matter of days. Monitoring these sudden shifts is crucial. Your chickens can get stressed from abrupt changes, affecting their health and productivity. Keep an eye on weather forecasts and adjust coop ventilation accordingly. On warmer days, ensure proper airflow to keep your flock comfortable. During chilly spells, close vents to retain warmth without sacrificing ventilation. Keeping the coop properly ventilated helps prevent respiratory issues that can arise from poor air quality.

Ensure your chicken coop is equipped with adaptive ventilation systems that respond to changes in humidity. Consider installing automatic vent flaps or solar-powered windows, which open and close based on temperature fluctuations, to effortlessly sustain the ideal conditions for your chickens.

Spring and fall also herald the molting season—a time when chickens shed old feathers and grow new ones. This process is energy-intensive, and your flock will need extra protein to support feather regrowth. Consider supplementing their diet with high-protein snacks like mealworms or sunflower seeds. These tasty treats will help them through this demanding phase. During molting, chickens become more vulnerable and may seek out safe spaces away from the flock. Ensure there are quiet corners in the coop where they can retreat without being disturbed by more boisterous companions. Providing extra bedding and low-light areas can help them feel secure during this somewhat uncomfortable period.

Supplement molting diets with probiotics and vitamins, ensuring the new growth is robust and healthy. Linseed and flaxseed can offer omega-3s for improved feather quality, supporting radiant plumage post-molt.

Parasite prevention becomes especially important during these transitional seasons. As temperatures change, parasites become more active, seeking out hosts in your flock. Regular coop cleaning can disrupt their life cycles, making your chickens less attractive targets. Scrub down surfaces and change bedding frequently to reduce parasite load. Natural deterrents like diatomaceous earth can also help. Sprinkle it in nesting boxes and along coop borders as a first line of defense against mites and lice. Keeping herbs such as lavender or peppermint near the coop can act as natural repellents, enhancing both scent and health within your chickens' living quarters.

Incorporate natural insect interceptors, like predator nematodes, into the soil around your coop to target larvae before they reach adulthood. Encouraging beneficial insects aids in keeping the parasite population under control naturally.

Springtime often brings a surge in egg production, and you'll need to prepare for this bounty. Consider adding extra nesting boxes to accommodate the increased output. This prevents overcrowding in existing boxes, reducing stress for your hens. Enhanced nutrition is another key factor during this period. Laying hens require additional calcium and nutrients to produce strong eggshells consistently. Layer feed supplemented with oyster shells or crushed eggshells provides the necessary boost. Checking water schedules is essential during increased egg production, as more eggs mean more water needed for your flock's sustenance and overall health.

Utilize crushed limestone as a regular supplement for calcium to aid in egg production. Introducing crushed eggshells back into your chickens' diet closes the cycle of use and replenishment sustainably.

Balancing these seasonal changes requires an eye for detail and a bit of patience. But with thoughtful preparation, you can ensure your flock thrives through spring and fall's ups and downs. Your chickens will reward you with robust health and plentiful eggs as they adapt to the rhythms of nature's cycles. As you observe their behaviors and needs, you'll find that these transitional periods are opportunities for growth—for both you and your flock.

The beauty of chicken-keeping lies in its ever-changing nature, with each season bringing new challenges and joys. As you tweak routines and make adjustments, remember the satisfaction that comes from caring for your feathered friends. Whether it's watching them scratch happily in the first warm days of spring or seeing them settle contentedly into cozy nests as autumn leaves fall, each moment spent nurturing your flock deepens your connection to this rewarding endeavor.

PROTECTING CHICKENS FROM PREDATORS YEAR-ROUND

I know we've touched on predator awareness in earlier chapters, but its importance cannot be overstated, so it bears addressing again here. Specific threats can vary throughout the year, and the change in seasons offers an opportune time to reassess our security measures.

The arrival of spring is often heralded by the cheerful chirping of newborn chicks and the gentle rustling of leaves. Yet, it also brings a different crowd—foxes on the prowl. As spring unfolds, foxes become more active, driven by the need to feed their young. This uptick in activity means your chickens might be on their radar. Similarly, in the fall, migrating hawks take to the skies, posing another threat as they hunt for easy prey during their long journeys. Understanding these seasonal shifts in predator behavior is key to safeguarding your flock. You want to stay one step ahead, knowing when to tighten security around your coop.

Consider scheduling routine patrols during high-risk periods, especially at dawn and dusk when predators are most active. This vigilance can deter potential intrusions before they occur.

Reinforcing your chicken coop's defenses doesn't have to be a Herculean task. Think of it as giving your fortress a yearly tune-up. Check fences for wear and tear, ensuring they're strong enough to withstand a determined predator's attempts to breach them. Repair any gaps or weak spots promptly. Installing secure locks on coop doors is another must-do. Raccoons are notorious for their dexterity, and a simple latch won't deter them. Opt for a lock that requires a bit of human ingenuity—a combination lock or a carabiner clip can work wonders.

Install predator-proof wiring that extends several inches into the ground around the coop. This barrier technique helps prevent digging predators, such as foxes, from tunneling their way in.

Sometimes, deterrence is the best form of defense. Motion-activated lights or alarms can startle would-be predators, sending them scurrying back into the shadows. The sudden burst of light or noise can convince them that your coop isn't worth the trouble. For aerial threats like hawks, consider adding predator-proof netting over your run. It's like putting up an invisible shield that keeps your chickens safe from above. This netting is sturdy yet unobtrusive, allowing sunlight to filter through while keeping danger out. Reflective metallic tape can also be strategically placed to deter predators with its startling, shifting light patterns.

Embrace technology by installing smart home cameras around the coop. These can alert you directly through your phone of any suspicious activity, allowing for rapid response and peace of mind irrespective of where you are.

Creating a community of fellow chicken keepers can be an invaluable resource in your quest to protect your flock. Networking with local chicken enthusiasts not only provides support but also creates a shared defense network. Keeping each other informed about recent predator sightings in your area can be a game-changer. You might hear about a neighbor who spotted a fox near their coop or learn that a hawk has been circling overhead. This information allows everyone to stay vigilant and adjust their strategies accordingly. Hosting a quarterly meet-up to share experiences and tips can provide fresh insights and foster a supportive environment for addressing predator issues collectively.

Community initiatives such as neighborhood watch groups or shared patrolling schedules further solidify a defense strategy, spreading knowledge and offering protection through collective effort.

PREDATOR DEFENSE CHECKLIST

- **Seasonal Awareness**: Stay alert to fox activity in spring and hawk migrations in fall.
- **Barrier Enhancements**: Regularly repair fences; install secure locks on coop doors.
- **Deterrents**: Use motion-activated lights/alarms; install predator-proof netting and reflective deterrents.
- **Community Support**: Network with local keepers; share predator sightings and tips.

This checklist serves as a quick reference to bolster your flock's security year-round. Through mindful preparation and community collaboration, you can create a formidable defense against the ever-present threat of predators. Recognizing the patterns of nature and adjusting your approach accordingly ensures that your chickens remain safe, allowing you to continue enjoying the many rewards of backyard chicken-keeping. With these strategies in place, you're not just reacting to threats—you're proactively creating an environment where your flock can thrive in peace.

Interwoven with the rhythms of the seasons are challenges that demand vigilance and creativity. Each step you take to protect your chickens reflects your commitment to their safety and well-being. The satisfaction of watching them scratch and peck without fear is a testament to the efforts you've invested in their care. In this dance with nature, your role as protector is ever-evolving, adapting to the changes with resilience and resourcefulness.

ADAPTING COOPS FOR SEASONAL CHANGES

As the seasons change, so does the environment around us. Your chickens feel it too, and adapting your coop to meet these shifts can make a world of difference. Imagine your coop as a living space that needs a wardrobe change with each season. Removable panels are a brilliant solution for ventilation control. During the summer, these panels can be removed to allow fresh air to flow through, helping to keep the interior cool. In winter, simply snap them back into place to keep the chill out. This flexibility means you're always ready for whatever Mother Nature throws your way.

Explore using magnetic or easy-slide panels for swift seasonal adjustments. These innovative designs cut down on time and effort while maintaining efficient changeover.

Seasonal roofing adjustments are another way to keep your chickens comfortable year-round. In the rainy months, consider adding a sloped roof extension or even temporary awnings to direct water away from the coop. This prevents leaks and keeps things dry inside. When the sun's blazing in summer, a reflective roof coating can help deflect heat, keeping the coop cooler. These tweaks are like giving your coop a seasonal makeover, ensuring it's always prepared for the weather ahead.

Additionally, consider installing retractable roofs or vents that can adjust based on moisture sensors, optimizing ventilation during downpours or heatwaves automatically.

Managing the coop climate effectively is all about balance. During the colder months, using heaters can provide much-needed warmth, but it's crucial to ensure they are safe and don't pose a fire risk. Conversely, fans can circulate air during warmer times, preventing overheating and reducing humidity buildup. Adjusting insulation is also essential for moderating temperatures. You might find that adding or removing layers of insulation as seasons change helps maintain a steady internal environment, reducing stress on your chickens.

Integrate solar-powered climate control solutions that adjust heating and cooling systems automatically. These self-sufficient technologies not only enhance comfort but reduce reliance on the grid, promoting an eco-friendly setup.

Routine maintenance is your secret weapon in keeping the coop running smoothly through all seasons. Regular inspections can catch small issues before they become big problems. As autumn leaves fall and spring showers hit, check for leaks or drafts that may have appeared. Cleaning gutters and ensuring drainage systems are clear will keep water from pooling around the coop— nobody likes soggy feet, least of all your chickens! These simple checks and repairs are like giving your coop a seasonal health check-up.

Schedule monthly cleanup sessions that involve thorough cleaning and checks of moving parts or flexible components, ensuring that all mechanisms operate smoothly without lag or obstruction.

Creative design solutions can make these seasonal adjustments easier and more effective. Convertible coop designs offer multi-seasonal functionality with minimal effort. Imagine a coop with sections that can be easily reconfigured to suit different weather conditions. It's like having a Swiss Army knife of coops—versatile and ready for anything! Whether it's adding a removable sunshade in summer or swapping out heavier blankets for lighter ones as spring arrives, these innovations make managing your flock easier.

Engage in modular design experiments that allow for customized coop setups, meeting diverse needs without over-hauling existing infrastructure. Tailored solutions ensure an ideal setup every season.

Inspiration can come from anywhere when it comes to coop design. Perhaps you've seen a neighbor's clever setup or stumbled across an innovative idea online. Don't be afraid to experiment and make your coop uniquely yours. Adding solar panels for lighting or rain barrels for water collection are great ways to enhance sustainability while catering to your flock's needs.

Remember that your chickens rely on you to create an environment where they can thrive regardless of the season. By making thoughtful adaptations and staying proactive with maintenance, you ensure their comfort and safety year-round. Plus, you'll find immense satisfaction in knowing you've created a well-oiled machine that serves both you and your feathered friends beautifully.

As you tweak and tinker with your coop, each change brings new learning experiences and opportunities for creativity. The beauty of adapting your setup lies in its ability to reflect not just the changing seasons but also your evolving understanding as a chicken keeper. Every adjustment is an act of care and commitment to your flock's well-being, showcasing the deep bond between you and these delightful creatures who share your space and life.

MANAGING SEASONAL BEHAVIOR CHANGES IN CHICKENS

As the days grow shorter and the sun dips below the horizon earlier, you might notice your chickens acting a bit differently. It's almost like they're signaling a shift in mood, much like how we feel a little off when summer turns to fall. One of the most noticeable changes is increased aggression during those shorter daylight hours. Your usually docile hens might become more territorial, vying for the best roosting spots or first dibs on feed. This behavior is often linked to the reduced light affecting their natural rhythms.

Individual tracking of aggressive behavior enables targeted interventions. Document specific cases with details on frequency and triggers, ensuring a tailored approach that minimizes disruptions.

Environmental stresses, such as sudden cold snaps or heatwaves, can also shake up the pecking order within your flock. When chickens feel under pressure, they often resort to reestablishing their social hierarchy as a way to cope. You may find that a previously dominant hen is suddenly challenged by a younger upstart. These shifts can cause temporary unrest, but with time and patience, things usually settle down.

To help manage your chickens' stress during periods of change or upheaval, consider implementing calming procedures that can significantly ease their temperaments. One effective method is to create a dimly lit environment, which can have a soothing effect on your flock. Additionally, the use of chicken-safe essential oils, such as lavender or chamomile, diluted in water and lightly misted in the coop, can also promote a sense of calm among your birds. Always ensure that any essential oils used are safe for chickens and applied in a manner that does not overwhelm them.

To help your flock navigate these changes, implementing stress-reduction techniques can be incredibly effective. Maintaining consistent routines and feeding schedules provides your chickens with a sense of stability. Think of it as sticking to a familiar routine when everything else seems unpredictable. Enrichment activities can also work wonders in keeping stress levels down. Scatter some scratch grains around the run or hang a cabbage for them to peck at —it's like giving them a puzzle to solve.

Incorporating interactive foraging systems into your chicken coop introduces a range of rotating challenges that are vital for both the mental and physical stimulation of your chickens. These systems, which feature various obstacles and feeding puzzles, are designed to replicate the natural foraging behaviors of chickens in the wild. They encourage your chickens to engage in activities like exploring, pecking, and scratching, mirroring their instinctual habits. This engagement is not only beneficial for keeping your chickens active but also plays a significant role in preventing boredom-induced conflicts within the flock. Implementing these enrichment strategies can be key to fostering a peaceful and healthy environment in your chicken coop.

During these times, adapting social management practices becomes crucial. If you notice one bird being particularly aggressive, it might be time to give her a little time-out. Separating aggressive birds temporarily can allow tensions to cool down without escalating into full-blown disputes. In the meantime, use distractions to reduce tension. A few strategically placed mirrors or shiny objects in the coop can captivate their attention and redirect energy away from pecking at each other.

For persistent aggression, explore integration training, focusing on socialization with unfamiliar environments or animals to reverse undesirable behaviors gradually.

Monitoring health and activity levels is equally important in managing seasonal behavior changes. Regular health checks help you catch any seasonal issues early. Look for signs of illness or injury that might be exacerbated by stress or weather changes—a droopy comb or a listless demeanor could indicate a problem. Keeping a record of behavioral observations allows you to spot patterns over time. It's like being a detective piecing together clues about your flock's well-being.

BEHAVIOR OBSERVATION JOURNAL

Document daily interactions and note any changes in behavior or social structure. Record incidents of aggression, health concerns, and responses to stress-reduction strategies.

This journal not only serves as a record but also helps you tailor your care approach based on your chickens' unique needs.

In this ever-changing dance with nature, understanding and responding to your chickens' behavioral cues fosters a harmonious environment. By recognizing shifts, reducing stress, and monitoring health, you are not only ensuring your flock remains healthy and content throughout the seasons but also gaining deeper insight into their unique personalities. You'll find that each chicken is an individual, as distinct as the changing seasons themselves, with their habits and quirks.

As we wrap up this chapter, remember that each season brings its own set of challenges and opportunities for growth—both for your chickens and for you as their caregiver. Embrace these changes with flexibility and patience, knowing that each adjustment strengthens your bond with your flock. In the next chapter, we'll explore how to seamlessly integrate your chickens into a sustainable lifestyle that benefits both them and the environment. As a guardian of your flock, your adaptability not only aids in their survival but enhances the quality of life for everyone under your care, enriching your understanding of nature's intricate and beautiful cycles.

CHAPTER 7
INTEGRATING CHICKENS INTO SUSTAINABLE LIVING

BENEFITS OF FREE-RANGE CHICKENS FOR SUSTAINABLE LIVING

E nvision a serene morning, enjoying the fresh air while watching your chickens roam and explore the backyard freely. Their feathers catch the light as they cluck and scratch, pecking at insects and seeds hidden in the grass. This is the essence of free-ranging, where chickens exhibit natural behaviors that promote their health and happiness. Allowing chickens to forage not only enriches their diet but also keeps them engaged and active. As they hunt for bugs and seeds, they exercise naturally, boosting their overall well-being. Let's not forget dust bathing— watching a chicken kick up dust as they roll around is a joy. This natural behavior keeps their feathers clean and their skin free from parasites.

Imagine the long-term effects of these natural habits. By engaging in natural foraging, chickens develop stronger immune systems, making them less susceptible to diseases, which consequently reduces the necessity for antibiotics or other medications.

Their robust health can lower potential veterinary costs and contribute positively to the overall ecosystem of your backyard. With each scratch and peck, they turn the soil, facilitating aeration, which over time, improves the garden's micro-environment.

Beyond their personal happiness, free-ranging chickens offer significant environmental benefits. By allowing chickens to roam, you reduce the need for manufactured feed, which cuts down on the resources needed to produce and transport commercial feed products. As they wander, they naturally fertilize the ground with their droppings, enriching the soil without synthetic fertilizers. This contributes to a more sustainable cycle, where waste becomes a resource rather than a problem. The chickens' droppings also improve soil structure, which can increase water retention and reduce erosion. This natural fertilization decreases the overall waste footprint of your backyard operation.

Consider the broader impacts on your community—when more people adopt free-range practices, the collective reduction in feed production and transportation could significantly lower carbon emissions. In essence, your choice to free-range chickens can ripple outwards, influencing broader environmental changes and fostering a culture of sustainability.

But the benefits don't stop at environmental impact; there's a nutritional advantage too. Eggs produced by free-range chickens often boast higher levels of omega-3 fatty acids, which are vital for heart health. The varied diet that free-ranging affords leads to eggs with richer yolk colors, indicating a better nutrient profile. These eggs not only taste better but also offer more nutritional value compared to those from confined hens. The meat from chickens raised on pasture can also be leaner and richer in flavor, making it a healthier choice for your table.

Picture a Sunday breakfast with the deep golden yolks of free-range eggs glistening on your plate. The satisfaction of knowing that these come not only from healthy chickens but also contribute positively to your health is unmatched. This fulfillment enhances your connection to your food sources, promoting a more conscientious and satisfying approach to eating.

Real-world examples abound in small-scale free-range operations showing increased egg production and improved chicken health. Consider a local farmer who switched from conventional to free-range and saw a boost in egg quality and output. Their chickens became more robust and lively, reflecting the positive impact of a natural diet and environment. Testimonials from other chicken keepers highlight similar stories: healthier flocks, tastier eggs, and a deeper connection to sustainable living practices. These successes illustrate how free-ranging can transform not just the quality of produce but also the joy of chicken keeping itself.

Engage with these examples and consider joining networks of free-range growers. Sharing experiences, successes, and challenges within these communities can provide support, inspire innovation, and reinforce your journey toward sustainable chicken-keeping.

Connect with local agricultural extension services or attend workshops to deepen your understanding of free-range techniques. These opportunities can broaden your perspective and introduce you to like-minded individuals dedicated to sustainable agriculture. Engaging with organizations that promote sustainable agriculture can provide a wealth of information and resources, further supporting your free-range endeavors.

USING CHICKENS IN PERMACULTURE AND GARDEN ECOSYSTEMS

The principles of permaculture revolve around creating sustainable systems that mimic natural ecosystems, and chickens fit perfectly into this vision. They act as natural pest control agents, eagerly seeking out insects that might otherwise plague your plants. By allowing chickens to roam among your garden beds, you enlist their help in managing pests without the need for chemical solutions. As they scratch and peck, they naturally aerate the soil, improving its structure and promoting nutrient flow. This symbiotic relationship enhances both the garden's vitality and the chickens' access to a varied diet.

Envision extending this dynamic partnership by incorporating chickens into your crop rotation plans. They can be moved into a plot after harvest where they help prepare the soil for the next planting season by eating leftover plant material and pests, turning your whole garden into a linked, self-sustaining system.

Incorporating chickens into garden spaces requires a thoughtful approach to protect your crops. Rotational grazing is a technique that prevents over-foraging by moving chickens between different garden sections. This method allows vegetation to recover while spreading the benefits of chicken activities throughout your plot. Portable fencing serves as an invaluable tool, granting you precise control over where your flock wanders. By redirecting their foraging paths, you ensure that your plants remain intact and thriving.

Strategically planting certain crops can create a mutually beneficial environment for both chickens and gardens. Comfrey, with its deep roots and high nutrient content, is an excellent choice. Chickens love to browse its leaves, and when it dies back, it adds valuable nutrients to the soil. Similarly, herbs like basil and mint repel pests that might otherwise damage your plants. By placing these aromatic herbs near vulnerable crops, you create a natural barrier against common garden pests.

Visualize the design of your garden layout with these considerations in mind—your garden could be a thriving ecosystem where every element supports the health of others. Sketch a map of your space to plan rotations and strategically place pest-repelling plants.

Imagine a garden where every element works together in harmony. Diagrams of integrated systems can help visualize how to achieve this balance. Picture a permaculture setup with raised beds for vegetables surrounded by pathways for chicken access. Portable fencing might section off areas for specific plantings, allowing chickens to roam freely without damaging delicate seedlings. Photos of thriving gardens can inspire you to design a space that suits your unique landscape while maximizing the benefits of chicken integration.

Creating a balanced ecosystem with chickens involves more than just letting them wander freely. Managing flock size ensures that your plant resources match chicken needs. Overgrazing can strip an area bare, so rotating their grazing areas helps maintain lush growth. Introducing biodiversity into your garden enhances resilience against pests and diseases.

Incorporate plant species native to your area to enhance your chicken-friendly garden ecosystem. These indigenous plants not only demand less maintenance but also bring significant environmental advantages. By choosing local flora, you create a natural habitat that supports a diverse population of beneficial insects and pollinators. These inhabitants play a crucial role in maintaining a balanced ecosystem by keeping pest populations under control, thus reducing the need for chemical pesticides. Furthermore, native plants are well-adapted to your garden's soil, moisture, and climate conditions, making them more resistant to diseases and pests. This resilience contributes to a healthier, more sustainable garden environment that supports your chickens in their foraging activities. They benefit from a richer and more diverse diet by having access to the wide variety of insects and plants that these native species attract, enhancing the nutritional value of their eggs and meat. Integrating indigenous plants into your garden not only supports your chickens' health and well-being but also promotes biodiversity and strengthens the local ecosystem.

Enhancing the habitat with shrubs and grasses offers chickens shelter and shade while adding visual interest to your garden.

Shrubs like elderberry or currant provide cover from predators and extreme weather while offering fruits for both you and your flock to enjoy. Grasses contribute grazing opportunities, creating a rich tapestry of textures and colors. Regular soil testing guides adjustments in planting strategies, ensuring optimal fertility year-round.

Diversifying your approach can significantly improve the efficiency and output of your chicken-keeping activities. Designating a specific section of your garden to create a tree guild represents a thoughtful strategy for fostering a balanced ecosystem.

A tree guild, at its core, is designed around a central tree chosen for its beneficial attributes, such as fruit or nut production, which serves as the anchor for the surrounding plant community. Around this key tree, companion plants are carefully selected and arranged in a way that they support each other's growth while maximizing the use of space. This could include smaller fruit or nut trees, berry bushes, beneficial herbs, and groundcover plants, each contributing to the guild in unique ways—some fix nitrogen in the soil, others attract beneficial insects, or provide mulch through leaf drop. Together, they create a symbiotic environment that requires less maintenance over time, as the plants' interdependent relationships reduce the need for fertilizers and pest control.

Incorporating a tree guild into your chicken-friendly garden not only enhances the biodiversity and resilience of your garden ecosystem but also offers an enriched foraging area for your chickens, promoting their health and well-being. This structure promotes a symbiotic relationship among plants, providing benefits like shade, shelter, and nutrients. By doing so, you make the most of limited space while fostering a supportive environment for your chickens and garden alike.

To involve your family in these activities, organize weekend projects like building a simple portable chicken coop together, incorporating fun challenges or games to make sustainability an engaging family affair.

Composting chicken manure is another cornerstone of integrating chickens into a sustainable ecosystem. This process transforms waste into valuable fertilizer by balancing nitrogen-rich manure with carbon materials like straw or leaves. Aerating compost piles accelerates decomposition, providing nutrient-rich soil amendments that enhance garden fertility. Chicken manure's high nitrogen content fuels leafy growth while improving soil structure and water retention.

Reducing waste through efficient practices further aligns chicken-keeping with sustainable living goals. Recycling chicken bedding as mulch or compost enriches soil while minimizing landfill contributions. Kitchen scraps become valuable feed, reducing food waste while supplementing poultry diets. Creative repurposing extends even to feathers, which can be crafted into insulation or art projects.

Innovative recycling methods highlight the potential for resourcefulness in chicken-keeping. Feathers can be transformed into crafts or used as insulation material. Chicken litter can undergo pyrolysis to create biochar, adding long-term carbon storage to the soil while enhancing fertility. These techniques exemplify how embracing sustainability fosters creativity and reduces environmental impact.

Engaging with local communities amplifies the benefits of sustainable chicken-keeping. Organizing coop tours fosters connections among fellow enthusiasts, sharing knowledge and inspiration. Workshops on sustainable practices empower others to adopt eco-friendly methods in their own backyards. Online platforms provide spaces for exchanging ideas and troubleshooting challenges together.

Collaborative projects strengthen community bonds while promoting resource sharing. Co-op purchasing reduces costs for feed and supplies while supporting local economies. Collective composting initiatives distribute resources efficiently, turning potential waste into nourishment for gardens across neigh-borhoods.

Success stories from communities thriving through shared efforts inspire others to join the movement toward sustainable living with chickens at its heart—whether through community gardens incorporating poultry or urban initiatives advocating for backyard flocks as part of resilient cityscapes.

Through integrating chickens into permaculture designs and community efforts alike, you contribute positively not only to your immediate surroundings but also toward broader ecological goals —transforming simple backyard endeavors into impactful steps toward sustainability on both local and global scales.

Leverage social media or local events to create a buzz around sustainable chicken-keeping practices, inviting people to explore and adapt these strategies to their contexts.

CREATING A BALANCED ECOSYSTEM WITH CHICKENS AND PLANTS

Watching chickens roam the yard, it struck me how much their presence transformed the landscape. They became part of a delicate balance, a dance between feathered foragers and the greens they care for so unintentionally. Maintaining this harmony is crucial. Managing flock size is a start; too many chickens can strip an area bare, leaving the ground barren and depleted. It's vital to match the number of chickens to the resources available, ensuring that plants have time to recover and thrive. Rotating chicken areas keeps them from overgrazing a single spot and allows vegetation to replenish. Think of it like crop rotation but with your flock, keeping both plants and chickens in prime condition.

Diversity should be embraced not just in plant and flower species but also in landscaping elements like ponds or rock gardens. These features attract helpful creatures like dragonflies or frogs, which further contribute to pest control, adding layers to your ecosystem's balance.

An ecosystem thrives on diversity, and introducing a variety of plant and animal life creates resilience. Native plants play a significant role in this balance, offering habitat support for local wildlife and beneficial insects. These insects, in turn, help control pests that might otherwise target your garden. By incorporating flowers that attract pollinators, you invite nature's helpers to your yard. The symbiotic relationship between chickens, plants, and insects forms a web of interactions that support ecosystem health. You'll find that a diverse garden isn't just more beautiful but also more robust against diseases and weather changes.

A wide variety of plants ensures that even if one species struggles, others can maintain the integrity of the ecosystem, proving the importance of biodiversity in resilient gardening and chicken-keeping.

Enhancing chicken habitats with plantings adds both beauty and function to your space. Consider shrubs like elderberry or lilac for shelter and shade. These not only provide your chickens with protection from the sun and predators but also bring vibrant color to your garden. Grasses such as rye or fescue offer grazing opportunities and ground cover. They're easy on the eyes and gentle on the soil, preventing erosion while feeding your flock. Creating these micro-habitats encourages chickens to explore and engage with their environment, making for happier birds.

Practical tips go a long way in keeping your ecosystem balanced. Regular soil testing can tell you a lot about fertility levels and whether your plants are getting the nutrients they need. It's a simple process but incredibly informative, allowing you to make informed decisions about what to plant where. Adjusting plantings based on seasonal changes is another key strategy. Some plants thrive in the cool of spring or fall, while others bask in summer's heat. By rotating crops and adjusting your chicken's access accordingly, you maintain soil health and plant vitality.

Consider setting up marker layouts or logs detailing which chickens access which sections. Over time, these records become indispensable in your decision-making, streamlining processes and informing strategies for upcoming seasons.

PRACTICAL TIP: SOIL TESTING FOR FERTILITY

To integrate soil testing into your garden management, procure a soil test kit from a nearby garden center or cooperative extension office. Follow the enclosed instructions meticulously to collect soil samples from different areas of your garden. The analysis will illuminate the nutrient content, pH levels, and any required soil amendments to enhance plant growth.

Balancing an ecosystem with chickens requires attentiveness and adaptability. It's about observing how each element interacts with others and making adjustments as needed. You might find that one area of your yard becomes a favorite dust bathing spot or that certain plants become targets for enthusiastic pecking. These observations guide your actions, helping you create an environment where all elements coexist harmoniously.

In this living tapestry, chickens play a vital role in maintaining ecological balance. Their natural behaviors contribute to soil health, pest control, and nutrient cycling. By understanding their impact and managing it thoughtfully, you foster an ecosystem that supports both plant and animal life. It's an ongoing process—a dynamic interaction that evolves with time and experience.

Observe how various areas of your garden weather different challenges. You may notice one section recovers quicker after heavy rain, indicating a potential for increasing rainwater harvesting opportunities in your chicken-keeping strategy.

With each season, you'll learn more about what works best in your space. Some years may bring surprises—unexpected pests or unusual weather patterns—but these challenges are opportunities for growth. By embracing diversity, enhancing habitats, and staying attuned to the needs of your garden and flock, you build a resilient system that thrives year after year. By reinforcing a connection between the garden and its chicken inhabitants, you form a more holistic approach to permaculture.

This approach turns chicken keeping into more than just a hobby; it becomes part of a larger commitment to sustainable living. It's about creating a space where chickens aren't just residents but integral members of a vibrant ecosystem. Whether you're new to this or have been at it for years, there's always more to discover and enjoy in this rewarding pursuit.

COMPOSTING WITH CHICKEN MANURE FOR GARDEN FERTILITY

Composting chicken manure is like turning a waste product into gardening gold. It's a straightforward process, but understanding the basics is crucial. Chicken manure is rich in nitrogen, which is fantastic for leafy growth, but it can be too potent if used directly on plants. The trick is balancing this nitrogen-rich manure with carbon materials like straw or dried leaves. Think of it as creating a perfect recipe, where ingredients must be carefully proportioned to achieve the best result. The carbon materials help to mellow out the strong effects of the nitrogen, making sure your plants get the nutrients without the burn. Aerating your compost pile is another key step in decomposition. It's much like fluffing a pillow—it keeps things light and allows oxygen to circulate, which helps microorganisms break down the material faster.

Involving family members in these processes can make composting a community effort, spreading enthusiasm and knowledge about the benefits of natural fertilizers among all age groups. Creating compost bins can be a family-friendly project where even youngsters learn the crucial part they play.

Chicken manure is a powerhouse for enriching garden soil. Its high nitrogen content fuels lush, green growth, making it ideal for leafy vegetables and other plants craving nitrogen. But that's not all. It also enhances soil structure by improving aeration and water retention. This means your garden will hold moisture better, reducing the need for frequent watering and increasing resilience during dry spells. Plus, composted chicken manure introduces beneficial microorganisms to the soil, boosting its overall health and fertility. These microbes are like little helpers working tirelessly to keep your garden thriving.

Further optimize your soil management by rotating the types of crops you grow, allowing your chicken manure compost to support varying nutrient demands while maintaining soil vitality year-round.

There are different methods to manage your manure, each offering unique benefits. Hot composting is one technique that speeds up the decomposition process. By maintaining a temperature between 140°F and 160°F, you can make nutrients available more quickly to your garden. This method requires regular turning and monitoring of moisture levels, but it's perfect if you're eager to enrich your soil fast. On the other hand, cold composting is less labor-intensive. It involves simply piling your materials and letting them break down over time, usually taking a year or so. While slower, this method requires minimal effort and still results in nutrient-rich compost.

Delving deeper into composting methods, consider vermicomposting, an innovative approach where worms play a crucial role in breaking down the manure. This method not only speeds up decomposition but also enhances the nutrient content of the compost. Worms process the manure, leaving behind castings that are rich in essential nutrients and beneficial microbes. These castings can be directly used in your garden, providing a quick and efficient nutrient boost to your plants. Another effective technique is trench composting, which involves burying layers of manure mixed with carbon-rich materials directly in your garden beds. Over time, these layers decompose, gradually enriching the soil with nutrients right where your plants need them. This method is particularly useful for gardeners looking to improve soil fertility without the need for a separate composting area. It's a straightforward, labor-saving approach that mimics natural soil-building processes, offering a practical way to recycle chicken manure directly into the garden.

When it comes to using composted chicken manure in your garden, timing is everything. Apply it during active growth stages when plants can most benefit from the nutrients. Early spring is often ideal for spreading compost before planting your crops. But be careful not to place it directly on plant roots as this can cause damage. Instead, mix it into the top layer of soil or use it as a side dressing around established plants. This approach ensures that nutrients slowly seep into the root zone without overwhelming your plants.

PRACTICAL TIP: TIMING COMPOST APPLICATIONS

To maximize the benefits of your compost, apply it in early spring or fall when the soil is warm and plants are either getting ready to grow or winding down their growth cycle. This timing allows nutrients to integrate into the soil before extreme weather conditions set in.

Using chicken manure as compost in your garden isn't just about recycling waste; it's about creating a sustainable cycle where everything contributes back to the earth. You'll find that with each application, your garden becomes more vibrant and productive. The soil's improved structure and fertility lead to healthier plants and higher yields, whether you're growing tomatoes or sunflowers. It's incredibly satisfying to watch your garden flourish from something that might have otherwise been discarded.

Consider starting small workshops or demonstration plots in your neighborhood to share these enriching practices with others, further nurturing a community committed to sustainability through practical actions.

Incorporating chicken manure into your gardening routine transforms what many see as a nuisance into an asset. With patience and care, you can harness its full potential and contribute to a thriving garden ecosystem. Your plants will thank you with bountiful harvests and robust growth, rewarding every effort you've put into managing this natural resource wisely.

Sharing your journey within local community gardens or social groups can amplify the collective impact, helping to spread awareness of sustainable practices and inviting others to contribute to the ecological well-being of the neighborhood.

REDUCING WASTE THROUGH EFFICIENT CHICKEN-KEEPING

Imagine standing in your backyard, watching your chickens peck contentedly at kitchen scraps tossed their way. This simple act is more than just feeding; it's a step toward reducing waste and promoting sustainability. Kitchen scraps, often destined for the trash, become a valuable resource for your chickens. They relish the variety, and you reduce waste production by incorporating these leftovers into their diet. Not only does this minimize food waste, but it also cuts down on the feed you purchase, saving money and resources. Recycling doesn't stop there; used chicken bedding can be a boon for your garden. Once it's spent in the coop, it transforms into mulch or compost, enriching your soil with nutrients that promote plant growth. This cycle of reuse and recycle illustrates how efficient chicken-keeping can significantly contribute to waste reduction.

When combined with strategies like composting, this feed replacement becomes part of a larger, harmonious cycle, bringing fresh perspectives on sustainable living into your daily routine.

Thinking outside the coop, there are innovative ways to repurpose chicken byproducts like feathers and litter. Feathers can be surprisingly versatile. As mentioned earlier, they can be used in crafts or as insulation material. Their lightweight structure makes them excellent for stuffing pillows or creating decorative items.

Meanwhile, chicken litter, commonly dismissed as mere waste, harbors untapped potential in the realm of biochar production. Chicken litter is the material that lines the floor of your chicken coop and run, and it plays a big role in keeping your flock healthy and your coop clean. At its most basic, chicken litter is a combination of bedding material (like wood shavings, straw, hay, or chopped leaves) and chicken droppings. Over time, this mixture builds up, and how you manage it depends on the system you're using—either regular cleaning or a method like the deep litter method.

The process of biochar production involves pyrolysis, where the litter is subjected to high temperatures in an anaerobic environment, effectively transforming it into biochar. This resultant form of charcoal serves as a powerful soil amendment, significantly enhancing soil fertility when incorporated into your garden beds. The addition of biochar to the soil does more than just dispose of waste efficiently; it also enriches the soil with stable carbon, which has a remarkable capacity to retain nutrients and water. This, in turn, boosts the soil's nutrient-holding capacity, leading to healthier plant growth and more bountiful yields. By integrating chicken litter into biochar production, you not only repurpose a readily available byproduct but also contribute to a cycle of sustainability that benefits your garden's ecosystem.

Conserving resources is a key aspect of sustainable chicken-keeping. Water conservation techniques are particularly crucial. Installing rainwater collection systems can provide a renewable water source for your flock, reducing reliance on municipal supplies. Gravity-fed waterers are another efficient option, allowing chickens to access water as they need it without waste. These systems ensure that every drop counts, especially during dry spells when water conservation becomes even more critical. Energy efficiency in coop design is also vital. Implementing solar panels or using natural light sources minimizes energy consumption while maintaining a comfortable environment for your chickens. Proper insulation reduces the need for artificial heating or cooling, keeping energy use low year-round.

Real-life examples highlight the success of these waste reduction practices. Consider a zero-waste chicken operation that utilizes every aspect of its setup to minimize impact on the environment. By recycling bedding, repurposing feathers, and conserving water and energy, these operations epitomize sustainable chicken-keeping. Testimonials from eco-conscious chicken keepers emphasize the benefits of such practices. One keeper shared how implementing rainwater collection saved hundreds of gallons annually, significantly reducing their environmental footprint. Another described the satisfaction of using biochar in their garden, witnessing improved soil health and plant growth as a direct result.

These stories underscore how reducing waste through efficient chicken-keeping aligns with broader sustainability goals. It's about creating a system where nothing is wasted and everything has a purpose. By embracing these practices, you contribute positively to both your local environment and global efforts to reduce waste and conserve resources. It's a rewarding endeavor that highlights the interconnectedness of chickens, gardens, and sustainable living.

Encourage neighborhood competitions or group activities focused on reducing waste and sharing results, transforming mundane waste-management tasks into fun, community-building initiatives.

These waste reduction strategies not only benefit the environment but also enhance your chicken-keeping experience by fostering a deeper connection with nature and sustainable practices. As you explore ways to minimize waste within your setup, remember that every small change can make a significant impact on both your backyard ecosystem and the planet at large. Embracing these practices means taking steps toward a future where chicken-keeping is not just about raising birds but also about nurturing a sustainable lifestyle that benefits everyone involved.

COMMUNITY ENGAGEMENT AND KNOWLEDGE SHARING

There's something incredibly fulfilling about knowing you're not alone in your chicken-keeping endeavors. Engaging with your local community can transform what might feel like a solitary hobby into a shared passion. Organizing chicken coop tours not only invites neighbors into your world but also opens up dialogue about sustainable practices. It's a chance for others to see firsthand how chickens contribute to a more eco-friendly lifestyle. Hosting workshops on sustainable chicken-keeping practices is another excellent way to foster community spirit. Whether it's teaching others how to build a predator-proof coop or sharing tips on composting chicken manure, these gatherings create bonds and spread knowledge.

Consider branching into local schools or community centers, organizing talks or clubs focused on chicken-keeping and sustainability—sparking curiosity and interest in upcoming generations.

The digital age offers endless opportunities for knowledge exchange. Online platforms like Backyard Chickens serve as vibrant forums where chicken keepers of all levels share advice, troubleshoot issues, and celebrate successes. Joining social media groups dedicated to poultry enthusiasts is like having a virtual coffee chat with friends who share your interests. These spaces are invaluable for finding answers to questions, learning about new trends, and connecting with a global community.

Collaborating on blog posts, joint YouTube videos, or cross-platform projects might open new pathways of knowledge sharing and engagement in your local networks.

Collaboration takes chicken-keeping to the next level. Imagine pooling resources with fellow enthusiasts for co-op purchasing of feed and supplies. Not only does this reduce costs, but it also fosters a sense of camaraderie. Collective composting initiatives are another brilliant idea. By working together, you can manage larger volumes of waste more efficiently, turning potential problems into garden gold.

Success stories abound in communities that embrace shared chicken-keeping efforts. Take, for instance, a neighborhood that transformed a neglected lot into a thriving community garden, complete with chickens. The project not only revitalized the space but also provided fresh produce and eggs for local families. In urban areas, initiatives promoting backyard chickens have gained momentum, proving that even city dwellers can enjoy the benefits of sustainable poultry keeping.

In urban projects, integrate elements like rooftop gardens, vertical plantings, or even tiny coops in balconies, adapting chicken-keeping to the urban context while further encouraging awareness and participation.

This communal approach to chicken-keeping enriches the experience for everyone involved. It creates networks of support, where each participant contributes knowledge and resources for mutual benefit. By engaging in these activities, you become part of a larger movement toward sustainable living. You're not just raising chickens; you're fostering connections and building a community dedicated to environmental stewardship.

Engage with local businesses or sponsors to create sponsorships for events, enhancing collaboration within community projects and bringing a broader awareness to sustainable practices.

CHAPTER CONCLUSION

As we wrap up this chapter on integrating chickens into sustainable living, it's clear that these feathered friends are more than just egg producers—they're catalysts for change. They bring people together, enrich our gardens, and contribute to healthier ecosystems. Whether you're connecting with neighbors or sharing tips online, remember that every step you take toward sustainability makes a difference. In the next chapter, we'll explore the intricacies of advanced chicken care techniques, diving deeper into understanding your flock's needs and behaviors. So, stay tuned as we continue this exciting adventure in chicken-keeping!

CHAPTER 8
ADVANCED TIPS AND TROUBLESHOOTING

HANDLING AGGRESSIVE CHICKENS AND MAINTAINING PEACE

The tranquil morning on a farm paints a picture of utopian peace where the countryside awakens with the gentle rise of the sun. The golden light stretches over fields and coops, promising a day of routine farming joys. Yet, within the confines of the chicken coop, a different drama might be unfolding —a chaos not of nature's making but chickens clashing out of sync with the placidity around. This volatile atmosphere among chickens can be as puzzling as it is unsettling. To restore peace, delving deeper into the causes of such aggression, alongside effective interventions, is paramount.

IDENTIFYING THE ROOTS OF AGGRESSION

Aggression often roots itself in scarcity of space, a common woe that transforms calm into conflict. Each chicken, like humans, harbors an intrinsic need for personal territory. When overcrowded like sardines in a can, their interactions can turn hostile. Drawing parallels to a contest where resources are limited, tension and conflict naturally escalate. In these tight quarters, a simple peck escalates rapidly into repeated brawls. To combat this, enhancing coop spaces or reevaluating flock numbers can alleviate these spatial pressures. Doing so grants each bird the needed breathing room, quelling stress-fueled aggression from the onset.

Furthermore, the disruption of a carefully maintained social hierarchy by introducing new members can lead to turmoil in an otherwise peaceful flock. Chickens, like many social animals, adhere to a pecking order that guides their interactions. Introducing new faces can topple this well-established hierarchy. Acting pre-emptively with measures such as temporary separations or quarantine periods before introducing newcomers can allow your flock to adjust, minimizing these social ripples.

STRATEGIES FOR CONFLICT RESOLUTION

Forging peace among chickens requires a strategic approach, blending gentle handling with thoughtful strategy. Temporary isolation stands as a potent method for defusing tensions among overly aggressive individuals. Remove the belligerents and allow the flock to settle into a rhythm that meshes peace over chaos. Reintroduce them gradually, ideally alongside treats or calm periods of interaction, to ensure smoother transitions and acceptance back into the pecking order without innate hostility.

Creating an engaging environment within the coop can also displace aggression. Stimulating chickens' curiosity sways their focus from conflict to exploration. Constructing a coop with varying levels and the inclusion of diverse structural elements can address boredom and aggression alike. Providing dust baths offers chickens not only a hygienic ritual but also a distraction, while perches facilitate natural behaviors, redirecting potential aggression.

Implementing behavioral modification tools and techniques could act as a supplementary aid for maintaining peace. The use of peepers or blindfolds to obscure a chicken's direct line of sight proves beneficial in reducing pecking tendencies, albeit appearing unconventional at first. These devices ensure the bird is not harmed, but rather aids in curbing aggressive behavior. This approach can be reinforced with positive reinforcement, rewarding peaceful behavior with treats, thereby nudging their default response towards calmness.

SUCCESS STORIES AND EXPERT INSIGHTS

Instances abound highlighting the transformations feasible with these methodologies. Consider Sarah, a diligent chicken keeper who battled ceaseless turmoil within her flock following the integration of new hens. Her toolkit included temporary separations, clever distractions like mirrors, and homemade hanging toys. These measures gradually dissolved tensions, fostering an environment of calmness. Key to her success was persistence and insightful observation, allowing her to tweak interventions as needed tailored to her unique environment.

Experts in the field echo the importance of deciphering the individual personalities present within a flock. Each bird possesses distinctive traits and quirks, urging a nuanced approach to their care. For instance, more assertive breeds might require stricter boundaries than their gentle counterparts. Maintaining a behavioral log for the flock can provide insights into patterns that necessitate intervention, thus cultivating a harmonious environment for all.

CASE STUDY: RESTORING HARMONY IN A TROUBLED FLOCK

Take Tom, who was confronted with an unruly rooster that disrupted the equilibrium of his flock. Through observation, Tom honed his skills in decoding behavioral cues, deciding to isolate the rooster temporarily while enhancing the coop environment with enriching activities. Dust baths and hanging treats transformed the once disharmonious setting into one of serenity.

Upon cautiously reintroducing the rooster to the flock, Tom took meticulous steps to ensure a smooth transition. He observed the interactions, ready to intervene at the first sign of aggression. Utilizing positive reinforcement techniques, such as offering treats during moments of calm behavior and gently guiding the rooster towards less dominant hens, Tom fostered a nurturing environment. This strategic approach enabled the rooster and the hens to gradually adjust to each other's presence, easing tensions and allowing the flock to organically reestablish its natural hierarchy and social structure.

By embedding these strategies within everyday practice, you nurture an environment where peace flourishes, to the benefit of both chickens and keeper. Each flock encapsulates its unique dynamics, fostering a need for keen observation and adaptability. Exercising patience, creativity, and empathy holds the key to overcoming challenges on your chicken-keeping adventure. Reflecting on the ongoing experiences and fine-tuning techniques as circumstances shift offer pathways to genuine harmony amid feathers and clucks.

CREATIVE SOLUTIONS FOR COMMON CHICKEN-KEEPING CHALLENGES

Venturing into your coop only to discover chickens pecking on their eggs as an impromptu brunch buffet is undoubtedly a frustrating scenario. Fortunately, inventive solutions abound for this conundrum. A fundamental yet effective measure is frequent egg collection throughout the day, which diminishes the opportunity for eggs to become targets of curiosity or hunger. Alternatively, installing roll-away nesting boxes ensures eggs safely roll into a protected compartment upon being laid, thereby keeping them out of chickens' reach.

Feather pecking presents another persistent challenge that transcends superficial annoyances to potentially cause stress and injury. Anti-pecking sprays, with their bitter taste, discourage such behavior by making the act undesirable. In severe cases, identifying the instigator allows for targeted interventions, damping the coop's stress levels. Engaging toys or mirrors furnish mental stimulation, aiding in alleviating boredom and reducing the urge to indulge in feather-pecking.

Noise, often a byproduct of chicken coops, may rile neighbors in proximity. Planting hedges around your yard creates a natural noise barrier, muting the clucks and crows that amplify in open spaces. To further contain sound, consider incorporating acoustic elements within the coop, such as insulated panels or acoustic foam. These materials can absorb sound without interfering with necessary ventilation, harmonizing the coop's acoustic landscape.

Maintain impeccable hygiene to uphold flock health; it needn't become a cumbersome task. Utilizing UV lamps can keep bacteria at bay on surfaces, serving as a chemical-free answer to sanitation. Scheduling deep-cleans that involve thorough bedding removal and surface disinfection prevents pathogen buildup. Eco-friendly disinfectants can be harnessed to advance hygiene goals with minimal adverse environmental effects.

INNOVATIVE COOP ENHANCEMENTS FOR IMPROVED FUNCTIONALITY

The farming realm is not left behind in the march of technological advancement. Automated feeder and watering systems afford ease by ensuring chickens' needs are met in your absence. These innovations can be tailored through timers or responsiveness, guaranteeing ready access to sustenance at all times. Moreover, smart lighting and temperature control systems introduce automated management of environmental factors, preserving egg production consistency and flock well-being.

Modular design elements ring in flexibility for chicken coops, featuring removable panels that allow straightforward access for maintenance. Expandable segments accommodate your flock's growth, thus assuring adequate space and comfort as bird numbers escalate. This adaptive framework ensures your coop's functionality persists despite changes in flock size while merging style with practical needs.

Incorporating sustainability in coop functionalities is increasingly paramount. Solar panels provide energy independence by powering lights or warming elements. Rainwater harvesting systems offer an eco-friendly method to supply water, easing dependency on municipal sources. Switching to renewable resources aligns practices with sustainability, enhancing the longevity and independence of your coop systems.

In the bustling environments of urban and suburban areas, where space is at a premium, innovative chicken coop designs have risen to the challenge. Vertical configurations extend the living spaces of chickens upwards, maximizing the limited ground area available. These coops are not only practical but also cleverly incorporate storage spaces for feed, tools, and supplies, ensuring everything needed for chicken care is neatly organized and easily accessible. Such architectural innovations seamlessly integrate into the smaller spaces of city living, offering a perfect blend of functionality and visual appeal.

Implementing these enhancements enriches not only the coop's functionality but also the chicken-keeping experience itself. The integration of technology, modular adaptability, sustainability, and creativity fosters a coop environment supporting both avian and human needs harmoniously. Whether rooted in countryside expanses or city confines, these creative approaches translate into a more efficient setup. The result is a flourishing haven—where hens lay and cluck serenely, offering the simple pleasures of fresh eggs and companionship secured in tranquility.

KEEP THE FLOCK GROWING

Now that you have everything you need to raise happy, healthy chickens, it's time to pass on what you've learned and help others get started, too.

By leaving your honest review of this book on Amazon, you'll be showing other new chicken-keepers where they can find the guidance they need—and helping them feel confident from day one.

Thank you for being part of this growing backyard chicken community.

Raising chickens stays strong when we share what we know—and you're helping me do just that.

👉 **Scan the QR code or go here to leave your review on Amazon:** https://www.amazon.com/review/review-your-purchases/?asin=B0FD478RY5

Thanks again for being here. I'm cheering you on—every step of the way!

— Avery Sage

CONCLUSION

As you reach the end of this guide, take a moment to reflect on the journey we've shared. From those first tentative steps into the world of chicken-keeping to mastering the nuances of flock management, you've come a long way. We began with the basics—choosing the right breeds, setting up a coop, and understanding local regulations. As you progressed, so did your knowledge and confidence. You delved into the intricacies of chicken behavior, health management, and even advanced techniques to create a thriving, sustainable ecosystem in your own backyard.

Throughout this book, my vision has been to empower you, a beginner, to not only raise chickens but to do so with joy and success. I wanted to equip you with the skills to gather fresh eggs and relish the satisfaction of sustainable living, all while connecting with nature. I hope you now feel equipped to achieve just that.

Let's revisit some key takeaways from our shared journey. You've learned how to recognize and respond to chicken behaviors, ensuring a harmonious flock. You've gained insights into managing health and wellness, from vaccinating to recognizing signs of stress and disease. You've discovered how to integrate chickens into a permaculture system, enriching both your garden and your flock's lives. Each chapter was designed to build your confidence and provide actionable steps to navigate any challenges you might face.

The benefits of keeping chickens are plentiful, and I hope you've felt the excitement of fresh eggs, straight from your coop. Beyond the eggs, though, you've embraced sustainable practices, reducing waste and enhancing your garden's fertility. And let's not overlook the personal satisfaction—the simple, profound joy of watching your chickens thrive, knowing you've played a part in their well-being.

As you move forward, carry with you the confidence that you can manage a healthy, productive flock. Remember, challenges will arise, but with the knowledge and tools you've gained here, you're well-prepared to tackle them head-on. Trust in your abilities, and know that mistakes are merely opportunities for learning and growth.

Now, it's time to take action. Whether you're just starting out or looking to enhance your current setup, use this book as your trusted guide. Begin by taking small steps, and gradually build on your achievements. Each step you take brings you closer to a flourishing flock and a more sustainable lifestyle.

I encourage you to engage with the broader chicken-keeping community. Join local groups or online forums to share your experiences and learn from fellow enthusiasts. These connections will enrich your journey, offering support and camaraderie as you continue to explore the world of chicken-keeping.

Reflect on your personal growth and accomplishments. Consider how far you've come since the beginning of this book. Each experience, each challenge overcome, has contributed to your development as a chicken keeper. Celebrate these milestones, and let them motivate you to keep growing.

Finally, I want to express my deepest gratitude. Thank you for placing your trust in this guide and for your commitment to this rewarding journey. It's been a privilege to accompany you, and I look forward to hearing about your successes. Here's to many happy mornings spent gathering fresh eggs and enjoying the delightful company of your chickens.

REFERENCES

1. camrynrabideau.com. (2024). *The Best Chicken Breeds For Beginners - Camryn Rabideau*. https://camrynrabideau.com/2024/03/27/the-best-chicken-breeds-for-beginners/

2. poultry.extension.org. (n.d.). *DEVELOPING REGULATIONS FOR KEEPING URBAN*. https://poultry.extension.org/articles/poultry-management/urban-poultry/developing-regulations-for-keeping-urban-chickens/

3. strombergschickens.com. (n.d.). *Essential Equipment For a Chicken Coop*. https://www.strombergschickens.com/blog/essential-equipment-for-a-chicken-coop/?srsltid=AfmBOopZEhrnf3wpGWdp5e YC0FT0VVuYN9OzB7ywinJub6s7BG6l6LVf

4. thespruce.com. (n.d.). *25 Free Chicken Coop Plans*. https://www.thespruce.com/free-chicken-coop-plans-1357113

5. lifeatcobblehillfarm.com. (2023). *Tips For Predator Proofing Your Chicken Coop*. https://www.lifeatcobblehillfarm.com/2023/09/tips-for-predator-proofing-your-chicken.html

6. blog.meyerhatchery.com. (2019). *Coop Ventilation and Why it is Important | Meyer Hatchery Blog*. https://blog.meyerhatchery.com/2019/12/ventilation-in-your-coop-and-why-it-is-important/#:~:text=Vents% 20placed%20high%20above%20your,prevent%20additional% 20moisture%20build%2Dup.

7. somerzby.com.au. (n.d.). *Chicken Coop Maintenance - A Step-by-Step Guide*. https://www.somerzby.com.au/blog/chicken-coop-maintenance/?srsltid=AfmBOor6BjTdVebea5gIlDhQDheMYgo-SBqOe71KY8EkhBQj6CweTTl_

8. merckvetmanual.com. (n.d.). *Nutritional Requirements of Poultry - Merck Veterinary Manual*. https://www.merckvetmanual.com/poultry/nutrition-and-management-poultry/nutritional-requirements-of-poultry

9. farmstandapp.com. (n.d.). *10 Best Organic Feeds for Raising Chickens That Promote ....* https://www.farmstandapp.com/6983/best-organic-feeds-for-raising-chickens/

10. cacklehatchery.com. (n.d.). *Safe Kitchen Scraps for Chickens - Cackle Hatchery*. https://www.cacklehatchery.com/safe-kitchen-scraps-for-

chickens/#:~:text=Soft%20items%20like%20cucumbers%20and,safe%20kitchen%20scraps%20for%20chickens.

11. flygrubs.com. (n.d.). *Seasonal Chicken Feeding Guide for Summer and Winter* .... https://flygrubs.com/blogs/news/seasonal-chicken-feeding-guide

12. merckvetmanual.com. (n.d.). *Common Infectious Diseases in Backyard Poultry.* https://www.merckvetmanual.com/exotic-and-laboratory-animals/backyard-poultry/common-infectious-diseases-in-backyard-poultry

13. backyardpoultry.iamcountryside.com. (n.d.). *3 Herbs to Heal and Prevent Chicken Respiratory Infections.* https://backyardpoultry.iamcountryside.com/feed-health/3-herbs-to-heal-and-prevent-chicken-respiratory-infections/

14. extension.psu.edu. (n.d.). *Deworming Backyard Poultry.* https://extension.psu.edu/deworming-backyard-poultry

15. merckvetmanual.com. (n.d.). *Vaccination of Backyard Poultry - Exotic and Laboratory Animals.* https://www.merckvetmanual.com/exotic-and-laboratory-animals/backyard-poultry/vaccination-of-backyard-poultry

16. ncbi.nlm.nih.gov. (n.d.). *Poultry - Effect of Environment on Nutrient Requirements* https://www.ncbi.nlm.nih.gov/books/NBK232332/

17. khpet.com. (n.d.). *10 of the Best Chicken Breeds for Eggs.* https://khpet.com/blogs/farm/10-of-the-best-chicken-breeds-for-eggs?srsltid=AfmBOopERpQYx4s7DmV_8PVnPWIGs3FOuy5NJj6ZM3TI4GF7HNTes3re

18. lalasfarm.com. (n.d.). *Chicken Molting: Nutritional Considerations, Potential* .... https://www.lalasfarm.com/post/chicken-molting-nutritional-considerations-potential-discomfort-and-insights-into-the-molt-season

19. farmhouseonboone.com. (n.d.). *Water Glassing Eggs - Farmhouse on Boone.* https://www.farmhouseonboone.com/water-glassing-eggs/#:~:text=Water%20glassing%20eggs%20involves%20submerging,day%20the%20hen%20laid%20them.

20. grubblyfarms.com. (n.d.). *5 Steps to Winterize Your Chicken Coop.* https://grubblyfarms.com/blogs/the-flyer/winterize-your-chicken-coop?srsltid=AfmBOoqMq1j-8EXHHRZyUCKAYJFGmIrKKWTFSPYZccaUKKVuzAXgK8QH

21. backyardpoultry.iamcountryside.com. (n.d.). *What is the Best Feed for Chickens in Summer?.* https://backyardpoultry.iamcountryside.com/feed-health/what-is-the-best-feed-for-chickens-in-summer/

22. the-chicken-chick.com. (n.d.). *11+ Tips for Predator-proofing Chickens*. https://the-chicken-chick.com/11-tips-for-predator-proofing-chickens/

23. chickenwhisperermagazine.com. (n.d.). *Keeping Chickens Comfy During Seasonal Changes*. https://chickenwhisperermagazine.com/the-chicken-movement/keeping-chickens-comfy-during-seasonal-changes/#:~:text=Sudden%20temperature%20changes%20can%20affect,signs%20indicate%20stress%20and%20illness.

24. strombergschickens.com. (n.d.). *Guide to Free-Range Farming: Facts, Pros, and Cons*. https://www.strombergschickens.com/guide-to-free-range-farming-facts-pros-and-cons/?srsltid=AfmBOoo61KfwMveU1wo_BlYitnBDAHqo3cbs_DLpFCpdYU5nnC_oUjha

25. thecritterdepot.com. (n.d.). *Integrating Chickens into Permaculture Gardens: A Guide ...*. https://www.thecritterdepot.com/blogs/news/integrating-chickens-into-permaculture-gardens-a-guide-for-sustainable-living

26. extension.unr.edu. (n.d.). *Using Chicken Manure Safely in Home Gardens and ...*. https://extension.unr.edu/publication.aspx?PubID=3028#:~:text=It%20should%20be%20composted%20or,around%20plants%2C%20people%20and%20pets.

27. econourish.co.uk. (n.d.). *10 Eco Tips for More Sustainable Chicken Keeping*. https://econourish.co.uk/sustainable-chicken-keeping-10-tips/

28. hobbyfarms.com. (n.d.). *How To Deal With Aggressive Chicken Behavior*. https://www.hobbyfarms.com/aggressive-chickens-behavior-tips/

29. backyardchickens.com. (n.d.). *Chicken Coops*. https://www.backyardchickens.com/articles/categories/chicken-coops.12/

30. gardenculturemagazine.com. (n.d.). *Permaculture Diaries: Why Raising Chickens in the Forest ...*. https://gardenculturemagazine.com/permaculture-diaries-why-raising-chickens-in-the-forest-is-best/

31. hobbyfarms.com. (n.d.). *These Apps Can Help You Manage Backyard Chickens*. https://www.hobbyfarms.com/these-apps-can-help-you-manage-backyard-chickens/

www.ingramcontent.com/pod-product-compliance
Lightning Source LLC
Chambersburg PA
CBHW031843200326
41597CB00012B/248